One World

JOHN C. POLKINGHORNE

One World

THE INTERACTION OF SCIENCE
AND THEOLOGY

TEMPLETON FOUNDATION PRESS
Philadelphia and London

Templeton Foundation Press
300 Conshohocken State Road, Suite 670
West Conshohocken, PA 19428
www.templetonpress.org

2007 Templeton Foundation Press Edition
Originally published by SPCK, 1986
© 1986 by John Polkinghorne
Preface © 2007 by John Polkinghorne

Designed and typeset by Kachergis Book Design

*Templeton Foundation Press helps intellectual leaders and others
learn about science research on aspects of realities, invisible and intangible.
Spiritual realities include unlimited love, accelerating creativity, worship,
and the benefits of purpose in persons and in the cosmos.*

The Scripture quotations in this publication are from the Revised Standard
Version of the Bible, copyrighted 1946, 1952, © 1971, 1973 by the Division of
Christian Education of the National Council of the Churches of
Christ in the USA, and are used by permission.

LIBRARY OF CONGRESS CATALOGING-IN-PUBLICATION DATA
Polkinghorne, J. C., 1930–
    One world : the interaction of science and theology / John C.
Polkinghorne.
        p. cm.
    Originally published: London : SPCK, 1986. With new pref.
    Includes bibliographical references (p.   ) and index.
    ISBN-13: 978-1-59947-111-2 (pbk. : alk. paper)   1. Religion and
science.   I. Title.
    BL240.3.P637 2007
    215—dc22
                                    2006026601

Printed in the United States of America

07   08   09   10   11   12   10   9   8   7   6   5   4   3   2   1

*To Reverend Mother and Sisters of the
Society of the Sacred Cross Tymawr Convent,
From a Priest-Associate*

For by a single impulse of God the Word everything is put together, each acting in its appropriate way, and all produce a single common order.

—St Athanasius, Against the Gentiles

Contents

Preface to the 2007 Edition

Apart from a short book intended to be a personal apologia for my Christian faith (Polkinghorne 1983), which I wrote during my transition from being a professor of mathematical physics to becoming an Anglican priest, *One World* was my first book on issues in science and religion, soon to be followed by two more (Polkinghorne 1988, 1989), with which it forms a trilogy. Reading it again, I recognize that it introduces a number of themes that have remained important in my thinking in subsequent explorations of the interaction between science and theology, those two great human engagements with reality.

I described the context of *One World* as being the post-Enlightenment realization that the quest for clear and certain ideas, which could serve as foundations for reliable knowledge, had, in the end, proved to be a heroic failure. In today's postmodern culture, with its questioning of metanarratives and doubting of claims of truthful understanding, the issue of what we can know and how we can gain knowledge is one of even greater criticality than it was in 1986. I believe that the answers to questions of this sort have to arise from the investigation of specific truth-seeking projects, assessing their actual methods and achievements, rather than from general epistemological argumentation. The consideration in this book of the nature of science in chapter 2, and of the nature of theology in chapter 3, are exercises of this kind. In both cases, we encounter communities that are seeking truthful understanding, which they affirm can be gained through the quest for well-motivated belief. The knowledge thus acquired cannot be asserted to be certain beyond any possibility of peradventure, but there are sufficient grounds for believing it to afford a reliable, if never complete, grasp of actual reality. In both science and theology, I believe that we can affirm a stance of critical realism—called "critical" because the method of enquiry is subtle, with a delicate intertwining of experience and interpretation, and called "realism" because the investigation has the character of a process of discovery, rather than that of human construction. This stance of critical realism is one commonly held by scientist-theologians like myself, and its defense is a topic to which I have often returned in subsequent writing (see particularly, Polkinghorne 1998, chs. 2 and 5). The fact that critical realism can be asserted of both science and theology, despite their very different subject material, implies a cousinly relationship between these two truth-seeking endeavors (Polkinghorne 2007), and it provides a basis for their mutual interaction as they present their different perspectives onto the one world of existent reality (see ch. 5).

Science advances at a rapid rate, and the chapter that would be most different were I writing the book today is chapter 4. What is discussed there still remains significant, and in this edition I have updated minor points, such as incorporating the recent revision of the age of the universe. Were I starting from scratch today, however, I would add more material. This would include an enlarged discussion of the anthropic fine-tuning of the universe, the remarkable particularity that has been recognized as necessary in the character of the laws of nature in order to permit the evolution of carbon-based life. Two excellent detailed analyses of this issue (Barrow and Tipler 1986; Leslie 1989) were published too late to be cited in *One World*. (For a concise but expanded discussion of this issue, see Polkinghorne 1991, 77–80.) The discovery of this striking specificity of our universe was an upsetting thought to many scientists, whose instinct is to favor generality over particularity. They would have preferred the cosmos to be just an unremarkable universe. If anthropic fine-tuning is to be made intelligible, it is certainly necessary to look beyond science itself, since the latter has to treat the form of the laws of nature as given, the unexplained basis on which it rests its explanations of specific happenings. One possible interpretation is theistic, seeing fine-tuning as an endowment given to creation by its Creator in order to enable cosmic fertility. Some scientists definitely disliked that option. To defuse the theistic threat, there has been an astonishing willingness to entertain the notion of a multiverse, the idea that our universe is simply part of a vast portfolio of different worlds, all distinct and disconnected from each other. Such ontological prodigality goes far beyond anything for which sober scientific motivation might be offered. It is a metascientific strategy of what seems to me to be a pretty desperate kind.

I am surprised that I made no mention in *One World* of chaos theory (for an account, see Gleick 1988), the discovery in the 1960s of intrinsic unpredictabilities present in the macroscopic physics of everyday processes, due to the existence of many systems of such exquisite sensitivity to the finest detail of their circumstances that their future behavior cannot be reliably foretold. Here is a scientific insight that can offer a metaphysical option for how one might begin to understand how the world might be open to its future. I have extensively pursued this matter as my contribution to the investigation into issues of divine action that occupied the science and religion community for much of the 1990s (for a survey of my ideas, see Polkinghorne 1998, ch. 3).

In the years since 1986 there has been continuing argument about evolutionary matters. Though much of this has centered on disputes about claims to discern Intelligent Design, I think that the most interesting scientific development has been Simon Conway-Morris's emphasis on the role of convergence in biological history (Conway-Morris 1998, 2003). It seems that certain kinds of structure often recur independently in biological evolution, indicating that the range of possibilities that are both functionally effective and biologically accessible is more limited than one might have supposed. Evolutionary history looks much less like a random lottery than many have claimed. Conway-Morris even believes that if there is extraterrestrial intelligent life, it is quite likely to be broadly similar to human life.

In relation to the issues discussed in chapter 6, there has been increasing interest, both scientifically and philosophically, in the processes by which new levels, such as life and consciousness, may be thought to have emerged in a context from which they were previously absent (see, for example, Gregersen 2003; Clayton 2004). This is a promising realm of enquiry, which one may hope will make further progress.

The pace of change in theology is less hective than in science. Development comes characteristically from deepening consideration, rather than from the discovery of completely novel concepts. In later writings I have returned to several of the topics briefly introduced in One World. They include the nature of scripture (Polkinghorne 1991, ch. 5; 2005, ch. 2), sacrament (Polkinghorne 2005, ch. 5), God's relation to time (Polkinghorne 2006, ch. 6), and the perplexing diversity of the world faith traditions (Polkinghorne 1996, ch. 10). A topic briefly referred to here (71–72; 91–92) is the eschatological hope of a destiny beyond death, both for human beings and for the universe itself. Recently it has figured as a significant issue on the agenda of science and theology, to which I have sought to contribute (Polkinghorne and Welker 2000; Polkinghorne 2002).

I have called my style of theological thinking "bottom-up," seeking to move from motivating experience to attained understanding in a way that is natural for a scientist, and which reflects the cousinly relationship that I see existing between science and theology (Polkinghorne 2007). In fact, when I wrote my Gifford Lectures (Polkinghorne 1996), they were subtitled "Theological Reflections of a Bottom-up Thinker," as I sought to address the central Christian beliefs expressed in the Nicene Creed in just such a fashion. I am a passionate believer in the unity of knowledge. *One World* is based on this conviction, whose truth I believe to be guaranteed by the unity of the one true God, the Creator of all reality.

Writing in the 1980s, I followed contemporary custom in using the male pronoun in reference to God. Of course, I was never so stupid as to think that God was gendered and actually male. I ask my readers today to excuse me for the faults of the past.

JOHN POLKINGHORNE
Queens' College, Cambridge
June 2006

Additional Bibliography

Barrow, J. D., and F. J.Tipler. *The Anthropic Cosmological Principle*. Oxford: Oxford University Press, 1986.

Clayton, P. *Mind and Emergence*. Oxford: Oxford University Press, 2004.

Conway-Morris, S. *The Crucible of Creation*. Oxford: Oxford University Press, 1998.

———. *Life's Solutions*. Cambridge: Cambridge University Press, 2003.

Gleick, J. *Chaos*. London: Heinemann, 1988.

Gregersen, N. H., ed. *From Complexity to Life*. Oxford: Oxford University Press, 2003.

Leslie, J. *Universes*. London: Routledge, 1989.

Polkinghorne, J. C. *The Way the World Is*. London: SPCK, 1983.

———. *Science and Creation*. London: SPCK, 1988; Philadelphia: Templeton Foundation Press, 2006.

———. *Science and Providence*. London: SPCK, 1989; Philadelphia: Templeton Foundation Press, 2005.

———. *Reason and Reality*. London: SPCK and Philadelphia: Trinity Press International, 1991.

———. *Science and Christian Belief*. London: SPCK, 1994; also published as *The Faith of a Physicist*. Minneapolis: Fortress, 1996.

———. *Belief in God in an Age of Science*. New Haven: Yale University Press, 1998.

———. *The God of Hope and the End of the World*. London: SPCK and New Haven: Yale University Press, 2002.

———. *Science and the Trinity*. London: SPCK and New Haven: Yale University Press, 2005.

————. *Exploring Reality.* London: SPCK and New Haven: Yale University Press, 2006.

————. *Quantum Physics and Theology.* London: SPCK and New Haven: Yale University Press, 2007.

Polkinghorne, J. C., and M. Welker, eds. *The End of the World and the Ends of God.* Harrisburg, PA: Trinity Press International, 2000.

Preface

My impression is that scientists are as likely to be religious believers as any other section of the community. Nevertheless, there is a feeling abroad that somehow science and religion are opposed to each other. Someone like myself, who is an Anglican priest and a (now honorary) Professor of Theoretical Physics, is sometimes regarded either with the amazement appropriate to the strange mixture of the centaur or with the wariness appropriate to the sleight-of-hand artist. Neither image is, I think, just. In fact, science and theology seem to me to have in common that they are both exploring aspects of reality. They are capable of mutual interaction which, though at times it is puzzling, can also be fruitful. This book is written to defend that thesis.

The undertaking is fraught with difficulty. Ideally it calls for someone expert in many aspects of modern scientific thought, in the philosophy of science and in theology. All I can claim is a degree of expertise in fundamental physics. I have long had an amateur interest in biology, philosophy of science, and theology, but, of course, I am no professional in these disciplines. However, I doubt whether there is anywhere an author competent to command the whole field. Moreover, if there were he would produce a book of an entirely different kind from that which I am attempting. It would be a summa of magisterial authority, but it would also by that very fact be inaccessible to most of the generally educated public who might be expected to take an interest in these matters. There is room, it seems to me, for accounts which might be called semi-popular; they aim to take the issues seriously but survey them in a way that is reliable without attempting the detailed argument that a professional discussion would demand. At their best such books can provide access to important general ideas and even prove modestly insightful. Writing of this kind is quite common in the scientific world (indeed I have written two such books myself) but less frequent elsewhere. Such an approach seems to me to be one suitable way of tackling some of the issues of science and theology. It is always open to the criticism of professionals who, rightly in their own domain, wish to dot each i and cross each t. The science, as much as the philosophy and theology, presented in this book is necessarily painted with a broad brush but not, I hope, inaccurately.

Much of the uneasiness felt in the mutual relationship of science and theology stems historically from the critical attitude of the Enlightenment. In important aspects we now live in a post-Enlightenment age, a point sketched in chapter 1. Part of the reassessment involved in the transition to the modern world has been the diverse accounts given of the scientific method itself. To a working scientist it often seems that the great men of the twentieth-century philosophy of science have tended to seize on one aspect of the complex scientific endeavor and then exaggerate it to the exclusion of balancing considerations. As so often, the best place to be seems to be in the middle, in this case adopting a cautious attitude of critical realism. Chapter 2 presents an outline of the issues. I am encouraged that W. H. Newton-Smith, in his professional account *The Rationality of Science,* reaches what seems to me to be a similar conclusion.

Scientists sometimes need convincing that theology is a rational activity, with phenomena for investigation and its own criteria by which to carry out that investigation. Chapter 3 attempts to present such a view of the subject. Obviously it can point only to one or two considerations, sufficient, I hope, to counteract the erroneous suspicion that all that is involved in theological discourse is ungrounded assertion. In fact, there is something encountered in religious experience which demands study and explanation. I write as a Christian, indeed as an Anglican, but most of the chapter is cast in very general terms. As a result there is more emphasis on personal experience and less on historical assessment than I would myself include if I were composing an apologia. I have attempted the latter exercise elsewhere (*The Way The World Is*).

The ground having been surveyed, chapter 4 begins the main substance of the book. It describes the worldview of contemporary science and illustrates why it is that the order, structure, and potentiality thereby revealed have encouraged many to feel that there is "more to the world than meets the eye." Science seems to raise questions which transcend its own ability to answer. Chapter 5 proceeds to a threefold discussion of the interaction of science and theology. It includes a defense of natural theology as an insightful (rather than demonstrative) discipline, actually practiced by many scientists in the face of the remarkable world that they investigate. This is preceded by a discussion of possible points of conflict (mainly concerned with our own and God's relationship to the physical world) and followed by a plea to use scientific ideas as an aid to the analogical imagination.

Any discussion of scientific and theological worldviews encounters the problem of reductionism. If in the end there were "nothing but" elementary particle physics, then the interaction of science and theology would simply be the swallowing up of the latter by the former. Chapter 6 outlines an antireductionist defense of level autonomy using physics itself as the principal source of insight. It also contains a discussion, necessarily partial and halting, of the small extent to which we can claim today to have some understanding of the interrelationship of mind and brain.

The closing short chapter presents the idea of sacrament as providing the point of intersection of scientific and theological understanding.

This book was conceived while I was Curate of St. Michael and

All Angels, Bedminster, and written after I became Vicar of St. Cosmus and St. Damian in the Blean. I am grateful to the people of both parishes for their friendly acceptance of my spending time in reading and writing. I also wish to thank my wife Ruth for reading the manuscript and helping to correct the proofs, and the editorial staff of SPCK for their help in preparing the manuscript for press.

JOHN POLKINGHORNE
Blean, Kent
September 1985

One World

1

The Post-Enlightenment World

The practice of religion and a belief in God appear to have been almost universal phenomena, with the exception of the modern Western world. We live in a pluralist society of belief and unbelief. That exception is often attributed to the rise of science. Our enhanced understanding of the physical world is held to have undermined the belief of many in a spiritual reality. Thus baldly stated, the proposition has the air of a logical non sequitur. Are not physical and spiritual matters concerned with different levels of meaning and thus no more in direct opposition than is the character of the page in front of you, discerned as composed of paper and ink, in conflict with the assertion that it is also the means by which its author is attempting to set his thoughts before you? The most comprehensive discussion of its chemical composition could not exclude consideration of its possible intellectual content. To see how science and theology have come to be thought of by many as being in some way in opposition requires a historical, rather than a logical, assessment.

Science first achieved a recognizably modern form in the seventeenth century. Those concerned in its early development were, almost all, people who took seriously the existence of a religious dimension to life. Many of the first Fellows of the Royal Society were of a puritan persuasion. Indeed, it has been suggested[1] that the Christian doctrine of creation, with its emphasis on the Creator's rationality (so that his

world was intelligible) and freedom (so that its nature had a contingent character which could be discovered only by investigation, rather
than by speculation) provided an essential matrix for the coming into
being of the scientific enterprise. Of course, some of those early scientists had problems in their relations with the ecclesiastical authorities
and with orthodox belief. We cannot feel that the Galileo affair reflects
much credit on the church, even if it was not quite the confrontation
between a lonely hero, imbued with a simple desire for truth, and
the forces of clerical obscurantism, that popular mythology depicts.
Newton had his difficulties in accepting Trinitarian belief, but neither
he nor Galileo was a skeptic, and Newton appears to have held the
mistaken view that his writings on the book of Daniel were of equal
importance to the *Principia*.

Yet in the seventeenth century we can also see the beginnings of
what was to become thought of as the conflict between science and
religion. Thomas Hobbes eagerly, if inexpertly, seized on the mechanical ideas then being developed to make the Democritean assertion
that reality was nothing but a concourse of atoms in motion. That was
a materialistic take-over bid, leaving no place for the mental, let alone
the religious. Few were inclined to be so dismissive of the claims of
mind to its own existence, but the manner in which those claims were
defended itself contained the seeds of future difficulty.

Descartes proclaimed the duality of mind and matter. How the
thinking substance of mind and the extended substance of matter
were related was not so easy to say. Ultimately he had to invoke God
as the guarantor of their connection. The Cartesian system, wittily
characterized by Gilbert Ryle as the ghost in the machine, had grave
problems. William Temple said that he considered that day on which
Descartes conceived it ("shut up alone in a stove"—a heated room—
as he tells us) was "the most disastrous day in the history of Europe."[2]
Such a judgment is the pardonable exaggeration of a philosophically
inclined person, but there is no doubt that Cartesianism had a dan-

gerous tendency. Its author's fondness for clear ideas drove him to a total divorce between mind and matter as the only tolerable way of preserving the claims of the former to equal consideration with the latter. Yet if too sharp a separation is made, the tangible nature of matter is liable eventually to promote a feeling of the unreality of intangible mind. Moreover, the Cartesian system is stiffly rationalistic. Clear ideas are excellent when we are able to conceive them, but it may be that at certain times with certain problems it is better to be content with a creative confusion than to strive for an oversimplified solution. Clarity can be purchased at the expense of the complexity of the truth. Even today our ignorance of matters germane to the relationship of brain and mind is such that we can proceed only with modesty and caution.

It was in the Enlightenment of the eighteenth century that the chill of mechanistic ideas communicated itself widely in a form less crude than that of Hobbes, and so more persuasive. The remarkable success of Newton's ideas in explaining the behavior of physical systems, both terrestrial and celestial, encouraged reliance on a discourse of reason whose paradigm was seen in the power of mathematics. That there might be aspects of reality, intuitively discerned, whose nature was fittingly expressed in the cloudier language of symbol was not taken sufficiently seriously. The thinkers of the Enlightenment sought by cold clear reason to comprehend an objective world of determinate order. They saw themselves as self-sufficient and were confident of their powers and of human perfectibility. Even theology was affected. When it did not lapse into a detached deistic belief in a God who had set the world a-spinning but cared little for it thereafter, it adopted a coolly rational tone, placing great reliance on natural theology's supposed demonstrations from the intricate design of the world. In line with the spirit of the age, God had become the divine Mechanic. There was considerable suspicion of religious experience less ordered and decorous than that provided by attendance at public worship.

The principal role of religion was thought to be the encouragement of morality. In England in the eighteenth century, one of the great theological figures was Bishop Joseph Butler, author of *The Analogy of Religion*. He warned John Wesley about the dangers of enthusiasm, saying: "Sir, pretending to extraordinary revelations and gifts of the Holy Ghost is a horrid thing, a very horrid thing."[3] Indeed, it is, if it is a matter of pretense, as Wesley agreed. But it is always possible that what is involved is, in fact, a valid experience which goes beyond the staid limits of conventional expectation.

Wesley and the other great preachers of the Evangelical Revival represent religion's protest against the frigid rationalism of the Enlightenment. Their preaching of human sinfulness may have been unhealthily guilt-ridden, but it took a more realistic view of the flawed condition of mankind than that provided by optimistic ideas of human perfectibility. Other protests were also made against prevailing rationalism. The poets and artists of the Romantic movement intuitively rejected the desiccated analytic method of the Enlightenment which, in Wordsworth's phrase, murdered to dissect. William Blake proclaimed with mystic intensity the preeminence of the symbolic over the scientific: "'What,' it will be questioned, 'when the sun rises, do you not see a round disc of fire somewhat like a guinea?' O no, no, I see an innumerable company of the heavenly host crying, 'Holy, Holy, Holy is the Lord God Almighty.'"[4] Blake's vision is powerful and disturbing but too idiosyncratic to have been widely influential.

In fact, as the nineteenth century progressed, the light of reason seemed to shine with ever greater clarity on a comprehensible and determinate world. Clerk Maxwell brought an order into the phenomena of electromagnetism which was fit to stand beside Newton's achievements in mechanics. The principles of physics seemed complete. All that was left was their application to problem solving. Above all, Darwin showed how competitive selection could sift favorable mutations from random variations, creating thereby the appearance of design without need for the intervention of a Designer. The one apparently convincing demonstration of the existence of God, on which the eighteenth-century theologians had placed such great reliance, was found to be fatally flawed. Paley had compared the likelihood of the intricate structure of the world being the result of chance to the assertion that a watch had been assembled by random causes. However, it now seemed that after all one could find a watch without a watchmaker's having had to put it there.

None of this logically denied the validity of religious experience or the existence of God. Yet it marginalized such claims in the minds of men. Like Laplace, whose demonstration of the inherent stability of the solar system made unnecessary Newton's belief in a divine corrective occasionally applied to stop the planets wobbling apart, people came to feel that they had no need of the hypothesis of God. The Enlightenment attitude had done its acid work, and many people's faith dissolved away.

By a curious irony, as the nineteenth century came to an end, the method and view of the Enlightenment were themselves beginning to dissolve in their turn. We now live in a post-Enlightenment age. The essential character of Enlightenment thinking was to allow the clear light of reason to play upon an objective and determinate world. Scarcely a feature of that description now survives intact.

The insights of depth psychology have modified our understanding of the operation of human reason. We are more than rational egos. The exact nature of the polarities within the psyche is a matter of dispute between Freud and Jung and their successors. Yet it is clear that our conscious minds are counterbalanced by an unconscious component, at once creative, chaotic, and teeming with symbol. These deep levels within ourselves need to be spoken to, and they themselves speak to and influence the ego of which we are aware. There is an element of Blake within all of us.

At the same time that the human psyche has revealed its shadowy and elusive depths, the physical world has denied determinate objectivity at its constituent roots. Heisenberg tells us concerning electrons and other elementary particles that if we know what they are doing we do not know where they are, and if we know where they are we do not know what they are doing. His uncertainty principle proclaims the unpicturability of the quantum world. Naive objectivity is a status inappropriate for its inhabitants. Moreover, the fitfulness inherent in quantum theory breaks the bonds of strict determinism. In general we can give relative probabilities only for differing possible outcomes of an experimental observation, and no cause is to be assigned for obtaining a particular result on a specific occasion. The world known to the twentieth century is a good deal more curious and more shadowy than the eighteenth and nineteenth centuries could have conceived.

That in itself is no great cause for religious rejoicing. The ancient Hebrews knew well the dangers of the waters of chaos. Our century has seen a recurrent cult of the Absurd which is destructive of true understanding. To acknowledge the limits of rationality, objectivity, and determinism is not to relinquish a belief in reason, a respect for reality, or a search for order. It may, however, lead to greater openness to the variety of the world and our experience of it, an acceptance that beside the insights of science, expressible in the quantitative language of mathematics, there are the equally necessary insights of religion, expressible in the qualitative language of symbol. It is the purpose of this book to explore and defend such a thesis. Our first task will be to evaluate the scientific enterprise itself, for it has not been immune from the threat of a twentieth-century dissolving process.

2

The Nature of Science

There is a popular account of the scientific enterprise which presents its method as surefire and its achievement as the inexorable establishment of certain truth. Experimental testing verifies or falsifies the proposals offered by theory. Matters are thus settled to lasting satisfaction; laws which never shall be broken are displayed for all to see. In actual fact, as we shall find out, the matter is a good deal subtler than that. Nevertheless, the great enhancement that the twentieth century has seen in our understanding of the world in which we live, even encompassing an account of its earliest moments fourteen thousand million years ago and including the beginnings of a comprehension of how life could have evolved from inanimate matter, together with the remarkable technological developments stemming from scientific advance, lends a certain credibility to this triumphalist point of view. Such splendid successes suggest that here is the key to real knowledge. In the bright light of science's achievements, other forms of discourse are in danger of appearing mere expressions of opinion. The widespread thought that science has somehow "disproved religion" is based on psychological effect rather than logical analysis. It is a continuation of the Enlightenment distrust of all knowledge which is not patterned according to the paradigm of scientific method.

It is ironic that at the same time that there is this widespread popular attitude there is also, in circles more austerely intellectual, a critical review of the nature of the scientific method and of its actual achievement. The practices of science have been reassessed and its procedures found to be more complex and questionable than the simple popular account acknowledges. The picture of the professor in his laboratory watching the pointer move across the scale to the expected reading, and thereby establishing his theory beyond the possibility of doubt, bears about as much relation to reality as does the simplicity of the comic-strip detective to the complexities of actual police investigation. If the method of science is open to revaluation, so, of course, will be the nature of the conclusions resulting from it. It is to these matters that we must now turn.

Certainly science seems to be successful in settling issues to the satisfaction of those concerned. At the beginning of this century, there were still a few physicists like Ernst Mach who did not believe in atoms. They thought the idea just a useful figment of the chemists' imaginations, but they did not accept the existence of a real granularity in nature. Nowadays, you would not find a scientist who would espouse such an anti-atomic opinion. We all believe in atoms, even if the elementary particle physicists think of them as "large" composite systems, and it is to the quarks and gluons that we now look for the basic constituents of matter. Such achieved agreement is impressive. It contrasts with many other forms of knowledge where debates continue without prospect of universal settlement. Karl Popper said, "But science is one of the very few human activities—perhaps the only one—in which errors are systematically criticized and fairly often, in time, corrected . . . in other fields there is change but rarely progress."[1] This apparent scientific progress is pretty clearly connected with the exploitation of the experimental method. Is it not science's power to manipulate and interrogate the material at its disposal which enables it to provide agreed answers to the questions raised? I write as a theoretical elementary particle physicist who for more than twenty-five years worked in that discipline. For the greater part of that period, the subject was experimentally led. It was the discoveries of our experimental colleagues which largely set the theoretical agenda.[2] Latterly theory has regained the initiative, with its development of gauge theories of fundamental interactions, but even so it is still to experiment that it must look for the confirmation of its ideas. In 1967 Abdus Salam and Steven Weinberg independently proposed an attractive theory which combined electromagnetism and the weak nuclear force (responsible for such effects as the beta decay of nuclei) into a single unified account. It was an idea comparable with the unification of the apparently dissimilar forces of electricity and magnetism which Clerk Maxwell had achieved in the nineteenth century. Nevertheless, Salam and Weinberg's work gained comparatively little attention until the confirmation in the second half of the 1970s of the existence of an effect that their theory had postulated, the so-called neutral current. Now the theory seems to us to be completely established because of the discovery in 1983 by physicists at CERN of the very heavy W and Z particles which are the cornerstone of its construction.

Told like that, it all sounds like a textbook example of the simple view that theory plus experimental verification equals established truth. Yet the story is a little bit more complicated in its details. The experimentalists could have discovered the neutral current in the 1960s, before Salam and Weinberg had formulated their ideas. They actually saw events which we would now understand as due to its effects. However, it is difficult to sift out this elusive phenomenon from the experiments, because many other things are also going on. Amongst this background (as the physicists call it) were events induced by neutrons which could look very much like those caused by a neutral current. To interpret the results, therefore, it was necessary to estimate how great these neutron effects would be. In the 1960s it was believed that these spurious background events would be sufficiently numerous to explain away altogether the apparent neutral current interactions. In the 1970s improved calculations showed that this was not the case. I do not doubt that the new calculations really were better than the old ones, but we have to recognize that people were motivated to do them partly by the theoretical expectation, by then existing, that there might well be an actual neutral current to be observed. This is just one of many possible examples of how difficult it is for an experimentalist to see what he is not looking for.

Furthermore, I have said that in 1983 physicists at CERN discovered the Ws and the Zs. Indeed, Rubbia and van der Meer were given a well-justified Nobel Prize in 1984 for doing so. What they and their colleagues actually saw, however, was a complicated pattern of readings in a very large and expensive array of electronic counters. An extensive chain of interpretation is necessary to translate those patterns into "Here we have a W," or "There is a Z."

The trouble with the simple view of scientific method is that it does not take into account the sophisticated web of interpretation and judgment involved in any experimental result of interest. To be told that the needle of the galvanometer moved to 7.6 on the scale, or that there are certain marks on a photographic plate, is not in itself a source of insight or excitement. It can only become so if, through careful assessment of possible competing and obscuring effects (the background calculations of which we have spoken) and from an interpretative point of view (which sets the result in a matrix of theoretical understanding and expectation), we are led to the conclusion that something of significance has actually happened. Experiments are always theory-laden. The dialogue between observation and comprehension is more subtle and mutually interactive than is represented by the simple confrontation of prediction and result.

In order scientifically to interrogate the world, we have to do so from a point of view. It is precisely this need for an (admittedly corrigible) theoretical expectation which distinguishes science from its precursor, natural history, which is simply content to take in the flux of apparent experience as it happens. In a famous phrase, Russell Hanson referred[3] to this theory-laden character of our observation as "the spectacles behind the eyes." Our scientific seeing is always "seeing as."

To recognize this is to raise the question of the character of our experimental knowledge. The role of observation as the stern and impartial arbiter of scientific theory is somewhat compromised if in fact the image of nature we receive is always refracted by those spectacles behind the eyes. Might there not be a variety of possible perspectives on the world of which the received scientific view at any time is just one option?

In books on the philosophy of science, this possible dilemma is often illustrated by the notorious duck/rabbit, a sketch which, looked at one way, can be seen as a duck and which looked at another way, can be seen as a rabbit, the open bill of one becoming the ears of the other. Actually, this particular ambiguity is rather readily resolved by acknowledging that what is before us is a rather exiguous line drawing. Physics itself provides a much more striking example of such ambivalence.

The conventional view of quantum theory,[4] accepted by the vast majority of physicists, states, for example, that there is no assignable cause for the decay of a radioactively unstable nucleus at any particular moment. All that can be asserted is that there is a calculable probability for such a decay taking place within a specified period of time. The quantum physicist is in the same practical position as the actuary of a large insurance company who is unable to say whether any particular client will die in the coming year, but who can be tolerably sure that a calculable number of clients in a particular age group will die within that period. However, there is an important difference between the physicist and the actuary, according to conventional quantum theory. There are causes why the actuary's clients die, even if they are not known to him. There are asserted to be no causes for individual events in the quantum world.

To this conventional quantum interpretation, there is an alternative point of view, first worked out successfully by David Bohm. It asserts that all events are causally determined, but some of these causes (called in the trade "hidden variables") are inaccessible to us. That is the reason, in Bohm's view, why our actual knowledge has to be statistical. It is a matter, not of principle, but of ignorance. This point of view is, of course, identical with that of the actuary, whose clients die of causes, to him unknown.

In the realm of non-relativistic quantum theory (that is, concerning the behavior of very small and slowly moving systems), the conventional theory and Bohm's theory give exactly the same experimental results. Yet the understandings they offer are radically different. Here is a duck/rabbit with a vengeance! Why then do the majority of physicists believe the one in preference to the other? It is clearly not a matter of observational decision.

I think there are two reasons for the majority preference for conventional quantum theory (which I share). The first is that Bohm's theory, though very clever and instructive, has a contrived air about it. It is significant that this is enough to put off most professionals despite the theory's "common sense" determinism, which might seem an overwhelmingly attractive feature to a layman. Matters of taste, judgments of elegance and economy, play an important part in the development of science. By these canons conventional quantum theory seems to most of us more elegant, and so more compelling, than Bohm's ingenious ideas. But why should the more elegant prove scientifically the more compelling, other things being experimentally equal? Here we see the coming into play of a factor, the search for simplicity, which goes beyond the impersonality of the popular account of the scientific enterprise. After all, is not one man's simplicity another man's complication? Does it not all depend on those spectacles behind the eyes? To Copernicus as much as to Ptolemy, the circle was the perfection of simplicity. It was only natural, in their view, that heavenly motion should be explained in circular terms. Kepler's introduction of ellipses must have seemed to many of his contemporaries a most ugly and unwelcome development. Simplicity only returned to celestial mechanics with the totally different beauty of the inverse square law inserted into Newtonian dynamics. Today we retain a belief in the elegance and economy of gravitational physics, though its current expression would be in terms of the geometrical curvature of space-time described by Einstein's general relativity (if one uses the language of classical physics) or in the gauge theory of massless gravitons (if one uses the language of quantum theory). Beauty is indeed in (or behind) the eye of the beholder. Its influence on scientific thought is undeniable, but that very statement raises the question of the true nature of that thought. A second reason for preferring conventional quantum theory to that of Bohm is that the former is much more readily combined with special relativity to give an account of small physical systems whose velocities approach that of light. Although many of its predictions are consistent with relativity, Bohm's theory requires a specific reference frame for its formulation. The contrasting success of conventional quantum theory in being readily open to this extension illustrates another, and reassuring, feature of "good" science—its fruitfulness, the way ideas can continue to be applied in circumstances going far beyond those for which they were originally invented. A curious twist to this part of the story is that no one has yet found a perfect reconciliation of conventional quantum theory with general relativity. In other words, despite great efforts, the quantum theory of gravity is not yet on a firm foundation, and the two modern points of view about gravitation sketched at the end of the preceding paragraph are not perfectly at one with each other. There is hope that the speculative theory of superstrings may provide the resolution of this dilemma. For the present it is interesting to note that physics can manage to survive with two of its fundamental theories, quantum theory and general relativity, imperfectly reconciled. Even pure theory is never exhaustively rational.

The simple account of science sees its activity as the operation of a methodological threshing machine in which the flail of experiment separates the grain of truth from the chaff of error. You turn the theoretic-experimental handle and out comes certain knowledge. The consideration of actual scientific practice reveals a more subtle activity in which the judgments of the participants are critically involved. If you wish to give an experimental physicist an uneasy moment, look him straight in the eyes and say, "Are you sure you have got the background right in your latest experiment?" (In other words, "Are you sure you have eliminated all possible sources of spurious effects and are actually measuring what you claim to measure?") If you wish to give a theoretical physicist an uneasy moment, look him straight in the eyes and say, "That latest theory of yours looks a little contrived to me." (In other words, "I do not see in it that look of elegant inevitability which time and again has proved the hallmark of true theoretical insight.") Their answers will not depend upon simple ineluctable prediction confronting indisputable fact. Rather, they will involve a reasoned discussion of how those concerned evaluate and interpret the situation.

This role of personal judgment in scientific work was emphasized by Michael Polanyi.[5] He called it tacit skill. Acts of discrimination are called for in concocting a successful scientific theory which are no more exhaustively specifiable than are the skills of a wine-taster in blending a good sherry. But just as the sherry blender has to submit the result of his labors to the judgment of the discerning public, so the scientist has to persuade his colleagues of the soundness of his judgment. This necessity saves personal knowledge from degenerating into mere idiosyncrasy.

Once one has acknowledged the part that personal discrimination has to play in scientific endeavor, the whole enterprise may seem to have become dangerously creaky, its rationality diminished or even destroyed, by the importation of acts of individual judgment, even if they are claimed to be validated by the eventual assent of the scientific community. Has not the austere search for truth degenerated into the proclamation of an ideology, even if democratically endorsed by its adherents? There have certainly been philosophers of science who have taken such a view, and it is from them that the scientific method has received its most severe criticism.

Thomas Kuhn studied those rare moments in the history of science when a major change occurs in the scientific worldview. Most of the time, scientists are engaged in problem solving, applying an agreed overall understanding to the attempt to explain particular phenomena. Just occasionally, however, it is the overall understanding itself which is subject to radical revision. An example of such a paradigm shift, as Kuhn calls it, would be the transition from classical to relativistic dynamics. For Newton there is a universal uniformly flowing time; for Einstein each observer experiences his own time so that two observers in relative motion will not agree about which events are simultaneous with each other. For Newton a particle's mass is an unchanging quantity; for Einstein it varies with the motion of the particle. Clearly there is a striking difference between these two systems of mechanics. We can all agree on that. But Kuhn proclaims a divorce between the two so absolute that he can say that "In a sense that I am unable to explicate further, the proponents of two competing paradigms practice their trades in different worlds."[6] This is his celebrated claim that two competing paradigms, such as Newtonian and Einsteinian mechanics, are incommensurable; that is, there is no point of contact and comparison between them. If this were really so, it would imply that there were also no rational grounds for preferring one to the other, since such grounds would depend on the possibility of making just such a critical comparison. Kuhn does not flinch from drawing that conclusion:

As in a political revolution, so in paradigm choice—there is no standard higher than the consent of the relevant community. To discover how scientific revolutions are effected, we shall therefore have to examine not only the impact of nature and of logic, but also the techniques of persuasive argumentation effective within the quite special groups that constitute the community of scientists.[7]

Thus Kuhn's study of scientific revolutions has led him to accentuate the role of the personal factor to the extraordinary extent of proclaiming the efficacy of scientific mob rule.

All this is really very curious and greatly overdone. Did special relativity really come to be adopted because Einstein had a propaganda machine superior to that of Lorentz? Experimental evidence (such as the eventual confirmation of the slowing of moving clocks via observation of the lifetimes of rapidly moving particles) presents perfectly adequate nonideological reasons for accepting the theory. While Newton's and Einstein's understandings of mass are very different, is there not sufficient residual common ground for us to be able to say that they are offering alternative, and so comparable, accounts of inertia? Kuhn dismisses as an irrelevancy the well-known fact that Newtonian mechanics is the slow-moving limit of Einstein's mechanics. Yet to physicists this relationship would seem to be important, for it explains why classical mechanics was so long an adequate theory and why it remains so for systems whose velocities are small compared with the velocity of light.

Of course, study of persuasive techniques can help us understand why a new scientific viewpoint gains quick or slow acceptance, but to suppose that this provides the major part of the story of how new ideas are embraced is surely preposterous. Indeed, in later writings

Kuhn himself seems to have withdrawn from so extreme a position. Kuhn's revolutionary incommensurability, if true, would undermine the idea that science can claim our rational, as opposed to rhetorical, assent. An even stronger threat to that idea is posed by the writings of Paul Feyerabend. He is a philosophical enfant terrible who does not hesitate to proclaim that the scientific emperor has no methodological clothes. Our discussion of skill has made it clear that there is no totally specifiable set of rules for scientific theory choice. There is no algorithmic machine, the turning of whose handle is guaranteed to lead to the Nobel Prize. At best, there are only guiding principles, exercised with discrimination by experts whose conclusions are subject to the collective judgment of the scientific community. Feyerabend seizes on this tacit, unspecifiable element and blows it up into a dominating principle of scientific laissez-faire. He claims that in science "the only principle that does not inhibit progress is *anything* goes."[8] He is a self-proclaimed scientific anarchist. What in Kuhn was simply preposterous becomes in Feyerabend the Theatre of the Absurd.

If science is an intellectual free-for-all, then there is no reason for preferring astronomy to astrology, the oxygen theory of combustion to the phlogiston theory. Feyerabend honestly recalls that "having listened to one of my anarchic sermons, Professor Wigner [a distinguished theoretical physicist] replied 'But surely you do not read all the manuscripts that people send you, but you throw most of them into the wastepaper basket.'" He acknowledges that he does so, but "partly because I can't be bothered to read what does not interest me . . . partly because I am convinced that Mankind, and even science, will profit from everyone doing his own thing."[9] His impishness is irrepressible.

Yet another assault on the rationality of science is mounted by adherents of what is called the "strong program" in the history of science. They assign to social forces a prime causative role in scientific change. For example, Andrew Pickering wrote of the recent sequence of investigations in high energy physics which have led physicists to believe that matter is composed of quarks and gluons, "The world of HEP [High Energy Physics] was *socially* produced."[10] The claim is that the largely unconscious adoption of certain conventions of experimental interpretation, together with a collective expectation framed in particular theoretical terms, has so molded the thought of the invisible college of high energy physicists that a quark model of matter was imposed on the supposedly plastic mass of available data. The assertion is there in the title of his book; it is "constructing" quarks, not "discovering" them. Weighty grounds would be required for so startling a conclusion. On investigation we find that all that is offered is an analysis of such incidents as the differing background calculations, which in the 1960s seemed to exclude neutral currents, but which in their revised form were the basis for confirming the by then theoretically acceptable neutral current in the 1970s (see p. 11). We can readily agree that this is an excellent example of how social forces can retard or accelerate the pace of scientific discovery, but there are no grounds at all for going on to assert that they actually control the nature of that discovery. After all, the dust does settle. No one would now claim that the neutral current is an artifact of background calculations. The cumulative weight of evidence for its existence, made clearer by increased understanding of how to perform the calculations of neutron-induced background, has simply settled that issue.

The wider question of the quark structure of matter is a more complex story, but it contains an incident which illustrates how it is that significant signals from nature provide the stimulus for the development of understanding. In the late 1960s it was discovered at Stanford that hard scattering took place from protons. That is to say, when projectiles such as electrons were hurled at protons, sometimes they bounced back. This was the direct analogue at the subnuclear level of Rutherford's famous work in 1911 when he found that alpha particles bounced back off atoms. He said later that the result was as surprising as if a 15-inch shell had recoiled on impact with a sheet of tissue paper. Rutherford interpreted his experiment as showing the existence of a pointlike concentration of positive electric charge within the atom. He had discovered the nucleus. In an exactly similar way, it was natural to interpret the Stanford hard scattering experiments as indicating the presence of pointlike constituents within the proton. From that time onward many of us felt sure that some form of quark model was an inevitable next development in particle physics. Of course, that conviction arose within an interpretative framework, as all scientific understanding has to do, but it arose in response to phenomena, not as an imposition upon them. If hard scattering had not taken place, no amount of social pressure would have succeeded in constructing a quark model.

The recognition of a role for judgment in the scientific enterprise, a tacit element not wholly reducible to the application of rules specifiable a priori, gives it a kindred character to aesthetic, ethical, and religious thinking. Many have asserted these latter modes of thought to be of a different and inferior kind, matters of mere opinion. We now see that what is involved in the comparison is a question of degree rather than an absolute distinction. To say so is not, as Kuhn and Feyerabend and others have suggested, to open the door to irrationality but simply to recognize that reason has a broader base than corresponds to a totally specifiable method of verification. (The qualification *total* is vital here; we are not saying that anything goes.) The mind has its reasons that computers know not of. The justification of this rational claim depends, I believe, on an assessment of the actual nature of the scientific achievement. By their fruits ye shall know them. We must consider what those fruits actually are.

At first sight the prospect might seem discouraging. Paradoxically, the advancing success of science appears subversive of its attainment of truth. Do not all theories in the end prove inadequate and have to be replaced? We once thought that the basic constituents of matter were atoms; then nuclei; then protons and neutrons; then quarks and gluons; next—maybe strings? Bigger fleas have lesser fleas, and so ad infinitum. Isn't ultimate scientific truth a will-o'-the-wisp? Newton-Smith calls this the pessimistic induction "any theory will be discovered to be false within, say, two hundred years of being propounded."[11] There is excellent evidence for adopting this maxim. However, to do so is only fatal if we thought that certain truth is our necessary goal. In fact we shall have to be content with the more modest aim of verisimilitude. Our understanding of the physical world will never be total, but it can become progressively more accurate. The analogy of a sequence of maps of increasingly larger scale may be helpful. None will ever tell us all there is to be told about that particular piece of terrain. Each is a kind of coarse-grained isomorphism, representing accurately features from a certain size upwards but ignoring or smoothing out those which are smaller. For different purposes different maps are adequate. As a motorist I do not need the detail I would demand as a hiker. In the same way established scientific theories do not disappear; they simply have their domain of applicability circumscribed. Newtonian mechanics is satisfactory for largish objects moving at ten miles an hour, unsatisfactory for the same objects moving at a hundred thousand miles a second. Scientific theories are corrigible, but the result is a tightening grasp of a never completely comprehended reality.

So I would wish to say. But there are many who would deny it. First, there is the problem of how we know, even within a prescribed domain, that we have arrived at an adequate map of its physical behavior. Newtonian mechanics has so far proved excellent for describing the collisions of billiard balls, but how can I be sure that the next time I approach the table they will not be found to be behaving differently? Anyone attempting to make a general statement faces the problem of induction, of how to produce universal laws from the study of specific instances. One cannot examine every electron in the universe before saying anything about electrons in general, nor can one survey every billiard ball collision that ever has been, or ever will be, before pronouncing on how such objects behave. That resolute skeptic, David Hume, was the first to emphasize the logical difficulty this presents. How then can science proceed?

One possible response is to moderate the claim. This is the attitude of Karl Popper. He exhibits a maximal distrust of induction. However many "for instances" there may be in favor of a theory, there is always an infinity of untried cases in which it might prove wrong. The odds are thus permanently stacked against its validity. In Popper's view, therefore, we abandon all hope of verification. The best that can be done is to settle for falsifiability. While any number of successes will never count in a theory's favor, one failure will prove fatal. "Only the falsity of the theory can be inferred from empirical evidence, and this inference is a purely deductive one."[12] This chilling message is conveyed in the original with the emphasis of italics. Clearly if that is all that can be said, the nature of the scientific enterprise is precarious indeed. "The empirical basis of objective science has then nothing 'absolute' about it. Science does not rest upon rock-bottom. The bold structure of its theories rises, as it were, above a swamp."[13] Popper is driven to this gloomy assessment because he exalts logic above intuition. (It is only the "deductive" which is safe.) Since the ratio of the number of successful answers to the number of potential questions is inevitably, for any theory, a finite number over infinity, those who rely on such wary calculation of odds will always condemn themselves to a state of intellectual pessimism. It is imposed by their timidity. Newton-Smith says "one cannot over-stress the counterintuitive character of [Popper's] position."[14] Is the Newtonian mechanics of billiard balls really in a state of permanent jeopardy? I think not.

Those who balk at induction do so because there is no exhaustively specifiable set of rules which enable one to lay down a priori when its application is justifiable. Its employment involves an act of judgment, even though in the case of a theory well tried in a definite domain, such as Newtonian mechanics, one cannot feel that great powers of discrimination are required for its successful exercise. We have already recognized that such acts of judgment enter other aspects of the scientific enterprise. That being so, an answer to the Humean criticism which is preferable to the partial surrender of Popper is simply to assert that we shall rely upon inductive method exercised with an appropriate degree of skill. Undoubtedly that attitude corresponds to the actual practice of science, and it seems to have stood the subject in good stead.

Science certainly appears successful. It has the air of progress about it. But what exactly is the nature of its achievement? Here we come to the second set of objections to any claim that science results in a tightening grasp of a never completely comprehended reality. It is asserted that the use of that last word is a naive misapprehension of what science is actually about. We have reached the parting of the ways between the positivists, the idealists, and the realists.

Those of a positivist persuasion lay stress on perceptions which can be intersubjectively agreed; the scientific task is the harmonization of such experience. Entities not directly accessible to experience, such as electromagnetic fields and quarks, which form the staple of the discourse of fundamental physics, are said to be just manners of speaking which are useful simply as means to that reconciling end. They do not represent actually existing realities. The scientific world is populated by pointers moving across scales and marks on photographic plates, rather than potentials or electrons; theories are just convenient summaries of data. There are, in fact, very considerable difficulties in drawing that clear distinction between the facts of data and the devices of theory which my simple summary of positivism has assumed.[15] The objections to positivism, however, go beyond that. Its arid account seems totally inadequate to explain the actual practice of science. After all, if all that happens is the reconciliation of various bits of experience, much of it recondite, why is it worth all the painful labor involved? Bernard d'Espagnat, speaking of the activity of elementary particle physicists, wrote:

Whereas the activity appears essential as long as we believe in the independent existence of fundamental laws which we can still hope to know better, it loses practically its whole motivation as soon as we believe that the sole object of the scientists is to make their impressions mutually consistent. These impressions are not of a kind that occur in our daily life. They are extremely special, are produced at great cost, and it is doubtful that the mere pleasure their harmony gives to a selected happy few is worth such large public expenditure.[16]

Or, I would add, the dedication and toil of those involved. The philosophical problems of positivism, together with its impoverished account of scientific motivation, mean that it has few adherents in its pure form. However, there are accounts of the nature of scientific achievement, less inadequate than positivism but substantially influenced by it, which make claims that fall short of the realism I am wishing to defend.

Science is concerned with the power to predict or the power to manipulate phenomena, we are told. These two abilities are closely connected, for to foresee is to be forewarned and so at least to some extent to be in a position to take action to obtain a desired outcome. Science does indeed manifest such instrumental capacity, but should we be content with that and not go on to claim that its final goal is understanding? An instrumentalist would maintain that the only question to ask about a theory is, "Does it work?" If it does, then we are not to bother whether it is true or not. The suggestion urged on Galileo by Cardinal Bellarmine, that the Copernican system was just a means of "saving the appearances" (of getting the answers right) but did not describe how things actually were, would be endorsed enthusiastically by someone of this persuasion. However, it will not do.

Suppose that the Meteorological Office was given a sealed machine which had the property that if you fed in details of today's weather, then

the machine would correctly predict the weather for any day ahead in
the following year. The predictive role of the Met Office in weather
forecasting would be perfectly fulfilled. Would that mean that all its
meteorologists would simply pack up and go home? Not at all! They
are also interested in understanding the way in which the earth's atmosphere and the sea and the landmasses interact as a giant heat engine
to produce our climate. Before long some of them would be tampering
with the seals on the machine in the hope of finding out how it worked,
expecting that that would lead to an improved comprehension of the
weather system that it modeled so accurately. No account of science is
adequate which does not take seriously this search for understanding,
together with the experience of discovery which vividly conveys to the
participants the impression that understanding is what they are actually attaining. I have never known anyone working in fundamental
physics who was not motivated by the desire to comprehend better the
way the world is. It is because they yield understanding, though often
having low or zero predictive power, that theories of origins, such as
cosmology or evolution, are rightly classed as parts of science.

To claim that understanding is the true goal of science and the
nature of its actual achievement is not of itself to have reached the
realist position I wish to defend. We have to ask the further question
of where this understanding comes from. Is it imposed by us, or is it
dictated by the nature of the world with which we interact?

The former account would be given by those who take an idealist
position. The modern grandfather of this point of view is Immanuel
Kant, who believed that space and time are necessary mental categories which we impose on the flux of experience in order to be able to
cope with it at all. This kind of view has not been without its support-
ers in the scientific community. Sir Arthur Eddington, in a famous
parable, compared physicists to fishermen using nets with a certain
width of mesh, who concluded that there were no fish in the sea
smaller than that particular size. In other words, the apparent ordered

reality that we think we perceive is alleged to be the product of our observational procedures. The American physicist Henry Margenau was bold enough to admit the consequences of such ideas and said, "I am perfectly willing to admit that reality does change as discovery proceeds."[17] In his view the neutron did not exist prior to its "discovery" in 1932. One's feeling that such a statement is, to say the least, highly unsatisfactory is reinforced by a consideration of the track record of idealist claims. Kant believed that he had demonstrated that space had to be three-dimensional Euclidean in structure. With our knowledge of non-Euclidean curved spaces, actually realized in general relativity, we can see that all that he succeeded in doing was to produce a specious rationalization of what at the time was thought to be the only physical possibility. Eddington spent the last years of his life developing the tortuous ideas published posthumously in his book *Fundamental Theory.*[18] Its supposedly rationally established conclusions have signally failed to correspond to the structure of the physical world revealed to subsequent investigation. If fruitfulness for the future is a good test of scientific creativity, idealist notions have proved a dismal failure.

We need not be surprised. The world, though ordered, is strange and subtle. Our powers of rational prevision are pretty myopic and limited by the contingency of the way things are, existing independently of how we think they ought to be. The natural convincing explanation of the success of science is that it is gaining a tightening grasp of an actual reality. The true goal of scientific endeavor is understanding of the structure of the physical world, an understanding which is never complete but ever capable of further improvement. The terms of that understanding are dictated by the way things are.

That is the realist position. It certainly corresponds to the way scientists themselves see their activity and are encouraged to persevere with it. Of course, most of them are philosophically unreflective people, and it might be that this is just a shared naive misapprehension.

Yet the way devotees of a subject view their practice must surely count for something in its evaluation. Many philosophers of science have been unwilling to give this due recognition, feeling that they knew best, without paying sufficient attention to what the honest toilers had to say. The realist view, it seems to me, is the only one adequate to scientific experience, carefully considered.

If realism is to prove defensible it has to be a critical, rather than a naive, realism. First, it has to recognize that at any particular moment verisimilitude is all that can be claimed as science's achievement—an adequate account of a circumscribed physical regime, a map good enough for some, but not for all, purposes. Once one moves outside regimes already explored, to hitherto unattained high energies for example, then there is every prospect that modification of our theories will be required to take account of unforeseeable phenomena. These modifications may, at times, be drastic (as when Einstein takes over from Newton), but there is sufficient residual continuity to discount the Kuhnian claim that we have lurched from one world to another, disjoint from it.

Second, our everyday notions of objectivity may prove insufficient as we move into regimes ever more remote from familiar experience. Quantum theory presents us with exactly this happening. According to Heisenberg's uncertainty principle, for entities like electrons we cannot know both where they are and what they are doing. This abolishes picturability in the quantum world. I shall discuss later (p. 53) in what sense we can still assert that an electron is "real," but it is clearly not that of naive objectivity. Realism is not tied to such simple notions derived from everyday experience alone.

Third, a critical realism is not blind to the role of judgment in the pursuit of science. It acknowledges that the simple picture of definite theoretical prediction confronting unquestionable experimental fact and leading to certain truth is too unsubtle an account of what science is about. As Newton-Smith says, "The story of SM [Scientific Method] will not produce a methodologist's stone capable of turning the dross of the laboratory into the gold of theoretical truth."[19] There are always unspecifiable discretionary elements involved.

We cannot take off our spectacles behind the eyes, but if experiments are theory-laden, it is also true, as Carnes points out, that theories are fact-laden.[20] They are responses to what is perceived to be there and in need of explanation. Perhaps the most troublesome question for the critical realist arises from the fact that for any finite set of data, there will always be a variety of possible theories which could fit it. (One could call this the duck/rabbit problem.) A rational criterion of choice is provided by demanding that an acceptable theory should prove its fruitfulness. It can do so in two ways: by a capacity to continue to cope with data as their range and accuracy expands, and by the theory being shown to have correct conclusions unforeseen at the time of its devising.

As an example of the former, consider the Newtonian account of the solar system. For about two centuries every new result coming from increased observational accuracy could be explained by a natural refinement of calculational technique. These theoretical responses represented fine-tuning in accuracy (for example, by taking into account hitherto neglected interplanetary effects) which was wholly in accord with the spirit of the theory and in no sense imposed upon it. (In contrast, a stubborn adherent of the Ptolemaic theory would have had to introduce ad hoc a new set of epicycles every time better observations were available.) The most striking illustration of such natural development of Newtonian ideas was provided by the work of Adams and Leverrier. They explained perturbations in the orbit of Uranus by supposing them to be due to a further, and till then unknown, planet. Their suggestions were triumphantly confirmed by the discovery of Neptune. The power of a theory to respond to progressive experimental probing without arbitrary manipulation is strong evidence of its verisimilitude. We cannot go on to say its truth, because its fruitfulness is not unlimited. An unresolved small discrepancy in the advance of the perihelion of Mercury eventually showed that even the Newtonian theory of gravitation had its limited domain of applicability. The explanation of this phenomenon required Einstein's general theory of relativity.

As an example of the second type of fruitfulness, we can consider Dirac's theory of the electron. In 1928 he devised an equation which successfully combined quantum mechanics with special relativity. Such a nontrivial synthesis was necessary to describe particles which are small and fast moving. It was an unexpected bonus when it was found that the same equation also explained the fact, till then mysterious, that the electron's magnetic properties were twice as strong as one would naively have expected. When this sort of thing happens, it is very convincing evidence for the verisimilitude of the theory. Again, it was no more than that, for it was eventually found that there are small corrections to the electron's magnetic behavior which require for their explanation the much more elaborate theory called quantum electrodynamics.

I believe that, after a certain time of testing, theories which gain wide acceptance in the scientific community have exhibited their reasonableness by demonstrating just such fruitfulness. Such rational staying power conveys an impression of naturalness and lack of contrivance which is convincing. Thus the underdetermination of theory by data does not pose a fatal difficulty for realism, since the theories which survive have been selected by the rational criterion of sustained success. Nor do I think that the lack of effective competing theories is to be attributed to a slothful acquiescence in a socially induced consensus. Scientists are active in a continual attempt to devise alternatives to received opinion, impelled not only by the search for truth but also by the desire to establish personal reputation.

I have attempted to defend a view of science which asserts its achievement to be a tightening grasp of an actual reality. In the course of the discussion, we have acknowledged the role that personal judgment, presented for the approval of the community and pursued along lines which are rational but not wholly specifiable, has to play in the enterprise. In my view this means that science is not different in kind from other kinds of human understanding involving evaluation by the knower, but only different in degree. It is clear that the personal element is less significant in science than in, say, judging the beauty of a painting, but it is not absent. We are to take what science tells us with great seriousness, but we are not to assign it an absolute superiority over other forms of knowledge so that they are neglected, relegated to the status of mere opinion. Our discussion has taken science off the pedestal of rational invulnerability and placed it in the arena of human discourse. It is not the only subject with something worth saying. If differing disciplines, such as science and theology, both have insights to offer concerning a question (the nature of humanity, for example), then each is to be listened to with respect at its appropriate level of discourse.

Finally, mathematics itself, the natural language of physical science, has not proved exempt from critical reassessment. This surprising development springs from Gödel's theorem.[21] This result asserts that in any mathematical system sufficiently complex to include arithmetic (that is, containing the whole numbers), there are propositions which are capable of being stated but not capable of being either proved or disproved. These undecidable propositions are, moreover, known not to be just pathological oddities but to include results of manifest significance. Instead of the completeness we would have expected from mathematics, it appears that every interesting mathematical system is open and incomplete. Because that is so, the consistency of mathematical systems becomes an incalculable question. Thus, even the exercise of mathematics involves an act of faith!

Hofstadter commented on this curious state of affairs, "Gödel showed that provability is a weaker notion than truth whatever axiomatic system is involved."[22] In other words, truth transcends theoremhood. Even in the austere discipline of mathematics, there is more than meets the calculating eye.

3

The Nature of Theology

Scientists often use the word "theological" in a pejorative sense, implying the absence of rigor and the presence of unmotivated assertion. Those who speak thus have a mental picture of theologians shutting their eyes and gritting their teeth in the effort to defend the indefensible, crying with Tertullian, "I believe because it is absurd." Paul Davies articulates a common view when he declares, "The true believer must stand by his faith whatever the evidence against it,"[1] and later he says, "Religion is founded on dogma and received wisdom which purports to represent immutable truth."[2] There is sufficient half-truth in these statements to make them dangerously misleading. If there is a God, he is a hidden God. He does not make himself known unambiguously in acts of transparent significance, invariably preserving those who trust him from every misfortune and regularly restraining and punishing the acts of transgressors. Neither prayer nor blasphemy is a magical lever which can be used to act upon God to make him demonstrate his existence. He is not to be put to the test,[3] either by the demand for a particular outcome or by challenge to his authority. If man has been given independence so that he may freely choose his response to God, then this elusive character seems necessary in One whose infinite presence, totally disclosed, would overwhelm our finite being. It means that religious experience has a mysteriously open character. The believer is ill and prays. If he recovers, he thanks God for his healing; if he does not, he seeks to accept that also as the will of God. Either way he believes he has received wholeness, given by the sustaining grace of God, whose exact nature is to be found only within the experience itself. The unbeliever may exclaim in exasperation, "Is God's head never on the block? Is it always 'heads he wins, tails you lose'?" The brilliant mistranslation of the Authorized Version does not accurately render the Hebrew, but it expresses exactly an element of the religious man's experience when it has Job say, "Though he slay me, yet will I trust in him."[4] To acknowledge this is to do justice to the character of what is involved in religious understanding, to recognize that it has its own nature which has to be respected. Yet in the appropriate terms, religious conviction is still to be evaluated. It may be impossible to lay down beforehand exactly what those terms are (just as we cannot give a priori specification of when the use of inductive arguments is scientifically justified). Only in the event itself can its meaning be found. Yet it is not the case that *whatever* the evidence that event yields, the believer will be uncritical in his assessment of it. The psalms contain many protests to God.

Tradition certainly plays an important part in religion. So it does in science. We inherit the legacy of those who have preceded us, and it would be disastrous if every generation had to start from scratch. I remember visiting a laboratory in Eastern Europe where Maxwell's equations (the fundamental equations of electromagnetic theory) were engraved on tablets of stone in conscious imitation of the Ten Commandments. Dogma (the root meaning of the word is "that which seems to be the case") and received wisdom are a necessary part of all human activity. The greater the role of personal judgment in a subject, the more we need the correctives afforded us by insights from the past. In that way we can best hope to allow for the tricks of intellectual perspective induced by the cultural conditioning of the present day. Science is least vulnerable in this way, and that is why its achievements present the cumulative character of increases in knowledge. The more personal the subject, the greater the risk that we are prisoners in the cultural cage of contemporary attitudes. The men of the past may have known things which are necessary but which we have lost sight of. For example, St. Augustine in the late fourth century understood the complex structure of the human psyche, with its internal polarities, in a way which has only recently been rediscovered through the insights of depth psychology. Moreover, science deals in generalities, in principle accessible to all; more personal forms of knowledge are also concerned with the illumination given to and through a particular individual at a particular time. Hence the so-called scandal of particularity, the emphasis on the unique. Our musical understanding and experience would be greatly impoverished if we balked at the scandal of the particularity of J. S. Bach and refused his musical offering conveyed to us in the tradition. In the same way religion looks to the insights of the spiritual masters. In the sphere of the personal, it is not inconceivable that the truest understanding of God is to be found in the possession of a wandering carpenter in a peripheral province of the Roman Empire, far away and long ago. Yet respect for the wisdom of other ages does not imply an idolatrous servitude to it. The Christian creeds are summaries of the church's insights into her experience of God, but each generation has to make them its own to the extent that it can. Theology, like science, is corrigible. There is nothing immutable in its pronouncements. If they are found wanting after careful investigation, then they are to be abandoned. Theology has long understood the distinction between truth and verisimilitude; every image of God is an idol which eventually has to be broken in the search for Reality.

The view of the theological enterprise which I would wish to defend is summed up in a splendid phrase of St. Anselm: *fides quaerens intellectum*, faith seeking understanding. Thus conceived, theology is reflection upon religious experience, the attempt to bring our rational and ordering faculties to bear upon a particular part of our interaction with the way things are. A. N. Whitehead wrote: "The dogmas of religion are the attempts to formulate in precise terms the truths disclosed to the religious experience of mankind. In exactly the same way the dogmas of physical science are the attempts to formulate in precise terms the truths discovered by the sense perceptions of mankind."[5] Two claims are being made here by Whitehead. One is that there is an analogy between the activities of theology and science, in that both are concerned with understanding and ordering experience. The other is that there is an identifiable sphere of human interaction with reality which can be attributed to "the religious experience of mankind." In other words, there is a natural source of raw material for the exercise of the theologian's art. Those who are not prepared to acknowledge this, or to take it seriously, will inevitably end up with an impoverished, or dismissive, view of the subject. People like Paul Davies, who say "I make no attempt to discuss religious experience or questions of morality"[6] have to recognize that in claiming nevertheless to speak of God, they resemble a cosmologist, prepared in framing his view of the universe to accept what he can see through a telescope, but refusing the extra information provided by radio- and x-ray-astronomy.

The fact of the matter is that there are widespread claims to the experience of a religious dimension to reality; of encounter with the numinous presence of an Other; the recognition of unity with a reality transcending oneself; the perception of a purpose at work in the world that carries the assurance (all things to the contrary notwithstanding) that all shall be well; the acknowledgment of an ultimate significance to be found in the way the world is. Of course, one must also recognize great variations in the manner in which this experience is apprehended and expressed. One of the major tasks of theology is to assess the extent to which the apparently conflicting claims of the world's religions can be understood as due to different culturally conditioned responses to the same reality, or the extent to which their seemingly canceling character indicates their reference is to fantasy rather than fact.

One of the strongest indicators of the validity of the claim that religion is in touch with reality is provided in the universal character of mystical experience, understood as the experience of unity with the ground of all being. It is to be found as an element in all the world's religions. William James concluded his survey of it by saying:

> This overcoming of the usual barriers between the individual and the Absolute is the great mystic achievement. In mystic states we become one with the Absolute and we become aware of our oneness. This is the everlasting and triumphant mystical tradition, hardly altered by differences of clime or creed. In Hinduism, in Neoplatonism, in Sufism, in Christian mysticism, in Whitmanism, we find the same recurring note, so there is about mystical utterances an eternal unanimity which ought to make a critic stop and think, and which brings it about that the mystical classics have, as has been said, neither birthday nor native land. Perpetually telling the unity of man with God, their speech antedates language, and they do not grow old.[7]

We can even, so to speak, perceive the rudiments of a mystical methodology, for it is by a stillness concentrating on a nondistractive point of attention (such as the act of breathing) that universally we are told that openness to the possibility of mystical experience can be enhanced.

Comparatively few people enter into an overwhelming act of union that can properly be called mystical. Yet there is considerable evidence that many people do have experiences less intense but recognizably religious in character. The biologist Sir Alister Hardy has been associated with a unit which has been collecting such material.[8] At present, the enterprise is in the "natural history" stage, the collection of anecdotal examples together with some surveys by questionnaire. Sufficient results have been accumulated to make it clear that the modern Western world is no stranger to religious experience, even if it is less likely to be associated with the explicit practice of one of the world's religions than would have been the case at other times and in other cultures.

One day as I was walking along Marylebone Road, I was suddenly seized with an extraordinary sense of great joy and exaltation, as though a marvelous beam of spiritual power shot through me linking me in a rapture with the world, the Universe, Life with a capital L, and all beings around me.[9]

Such moments of insight prove commoner than many people imagine. It is the task of theology to help us to assess their true nature and what they tell us about the world in which we live.

The appeal to religious experience characterized much nineteenth-century theological thinking, in reaction to the dry rationalism of the eighteenth century. The great German theologian Schleiermacher concentrated on the feeling of absolute dependence, seen as an indication that man is not autonomous but needs God for his completion. One of his successors, Ritschl, placed emphasis on morality and the assignment of value as signs of the irreducibility of religion to other experience. I believe that deeper mysteries than these are involved in our religious life, but at least such approaches ground theological thought in its own domain of human relevance.

The Anglican theological tradition in which I seek to stand perceives a three-fold basis for its enquiry. First, there is scripture, that is to say, the record of the great teachings, great events, and great figures of the past which we believe are of particular significance for us in our search for God and an understanding of his ways with men. For the Christian the focus of attention is the life, death, and resurrection of Jesus Christ, for we believe that in him God has disclosed himself in human terms. For accounts of Jesus, including his impact upon his immediate followers, we have to turn to the New Testament. Because he was a Jew, and his first followers largely Jewish, we also need to read the Old Testament record of God's dealings with Israel and her developing understanding of God. Without that background, assumed by the New Testament writers, much of what they have to say is only partly intelligible.

Writings which are treated as scripture have about them a special quality of authority. It is part of the religious experience of reading them to be prepared to submit oneself to what they have to say. It is in this mode of reading that we become aware of what is meant by speaking of their inspiration, their spiritual value. This is not, however, the only way to read these writings. They are also available for us to use as evidence by which we can, for example, seek to reach a critical assessment of the kind of man Jesus was, the sort of claims he made, the significance of his strange and humiliating death, and the truth or otherwise of the assertion that he was raised from the dead. When we read it thus, we do not treat the New Testament as an unquestionable authority; we submit it to the analysis of reason. We, recognize, for example, that the Gospels are not biographies in a modern manner but interpreted accounts of the things about Jesus preserved by those who honored him as their Lord. I have elsewhere[10] tried to sketch my own attempt at such an enquiry and how it leads me to affirm the Christian assertions about Jesus, including that of his resurrection. This use of scripture is closely akin to the handling of evidence in observational science. Just as a paleontologist gains insight from a specific part of the fossil record, or an astronomer finds that the unusual event of the regularly flashing pulsar affords confirmation of his ideas about neutron stars, so in the phenomena of the New Testament the Christian finds those particular sources of enlightenment on which he grounds his understanding of God and of the world. Special regimes are sources of particular understanding.

The second basis for theological enquiry is tradition, that is to say, the record of religious experience to which we add our own mite of personal knowledge. Our concern now is with the continuing interaction of God with humanity, rather than the unique testimony of the founding figures of a religion. The record of the latter in scripture will not continue to speak to successive generations unless it bears some consonant relation to what is experienced now, though it is also true that part of that present experience is the encounter with the spiritual testimony of scripture. In Christian theology the collective witness and insight of the church, the believing community, is an important part of the tradition. It does not, however, exhaust it. We must also take account of the experience of people of other religions, including their claim to possess writings of the nature of scripture, and the religious experience, or lack of it, of those who stand outside any historic religious tradition. There is much that Christians can learn, and need to learn, from others. Yet, if tradition is to have any meaning, there has also to be the core of our own personal experience. This is the nearest theology gets to having an experimental component. The analogy is not very close, though, because of the nature of religious experience, not subject to advance specification. This open character is the reason why symbol is the language of religion. Mathematics is precise, and so its use is appropriate to physical science with its power to put matters to the experimental test. To do so requires that we should be able to lay down clearly beforehand what our expectations are. In the personal encounter of religion, the meaning to be met with cannot be thus specified in advance; it is to be found only in the event itself. In this domain we need recourse to the symbolic. A symbol is always less than the reality to which it refers, just as a word is less than the concept it signifies. The openness of words to richness and variety of meaning, the cloud of allusion, is what makes poetry possible and a dictionary ultimately impossible. A great work of art, which is an articulation of symbol, is capable of providing many levels of encounter, and we all meet it in different ways. The gap between symbol and referent gives just that room for maneuver which is essential to describe the religious realm. Religion is concerned with personal encounter, not a packet of propositions. Commitment is involved. As Ian Barbour dryly observes, "God is not worshipped as a 'tentative hypothesis.'"[11]

The third basis for theological enquiry is reason. Not only must we exercise our rational faculties, but a concern for sound learning will encourage us to examine the relation of religious assertions to other assertions about the world and to assess the degree of consonance we find between these differing discourses. Subsequent chapters will be much concerned with this. For the present let us acknowledge that the use of reason does not exclude the possibility that certain types of subject matter call for their own particular mode of rational discourse. Reason is not a euphemism for an inappropriate tyranny of "common sense." A striking illustration of the need for broader views is to be found within science itself. You tell me that Bill is at home and that he is either drunk or sober. I conclude that either I shall find Bill at home drunk or I shall find him at home sober. The learned would say that I have applied the distributive law of logic to reach this simple conclusion. It seems pretty obvious, but in quantum theory a similar line of argument would in fact be fallacious.[12] In consequence there is a special quantum logic which differs from that known to Aristotle or the man in the street. If there is need to accommodate reason to the idiosyncratic nature of subatomic particles, may there not be need for even greater subtlety in exercising it on the nature of God?

The three-fold basis of scripture, tradition, and reason provides a public domain for theological discourse and delivers the discussion from the confines of an enclosed world of personal preference and idiosyncratic experience. When studying basic theology in preparation for ordination, I found that it was people like New Testament theologians who were most congenial to the thinking of a physicist. Their principal concern was with phenomena, what it was that called for explanation. When it became known that I was going to become a clergyman, my scientific colleagues had quizzed me about my Christian belief. In attempting to reply to them, I knew where I had to begin. The instinctive question was, "What is the evidence that makes you think this might be the case?" In half an hour's chat over a cup of coffee in some laboratory canteen, one could do no more than give the sketchiest of answers, just as one would be equally inadequate in similar circumstances in answering an arts colleague's enquiry about why one believed matter to be composed of quarks and gluons. In both cases there is a history of events and interpretation and critical thought which has to be surveyed. In the end I wrote a little book *(The Way The World Is)* setting out the apologia which time had not permitted me to make in that laboratory canteen. It is largely concerned with the appeal to historical events and their developing interpretation.

Many of the problems of speaking about God concern his relation to time. There is a school of theological thought, called process theology, which sees God as developing within time, realizing himself through the evolution of his creation. This is in sharp contrast with the classical Greek conception of God as static perfection inhabiting eternity, detached from the flux of the world.

Christian thinking has often been influenced by this Greek idea, but the God of the Bible is far from remote. He is involved in the world, but not tied to it to the extent that the process theologians suppose. He is indeed "the King of kings and Lord of lords, who alone has immortality and dwells in unapproachable light,"[13] but he also "looks far down upon the heavens and the earth." "He raises the poor from the dust, and lifts the needy from the ash heap."[14] A God who loves could never be content with the isolation of perfection, and if he has given humanity the freedom of choice, must it not be the case that he is open to the unforeseeable consequences of their decisions? The very nature of love implies vulnerability, the gift of freedom entails precariousness about the outcome.[15]

Modern physics associates space and time and, via the general theory of relativity, links both with matter, so that if the material world sprang into being at the fiery singularity of the big bang, then time itself began then also. "When" the world was not, there was no time. That remarkable man, St. Augustine, had reached a similar conclusion fifteen hundred years before Einstein.[16] There is no particular difficulty scientifically in envisaging God as outside space and time (in some super-dimensions if you wish to be prosaic in your imagining) "looking down" on the whole evolution of the universe laid out before him. However, so atemporal a deity is too close to the God of the Greek philosophers for Christian comfort. The God of Abraham, Isaac, and Jacob, who is active in the affairs of men, who suffers when his people suffer, cannot be so wholly above the struggle of life. While God is in his essential nature eternal and unchanging, his act of creation and his love for his creatures implies a self-limiting and self-emptying—a kenosis, as the theologians say—by which he allows the vulnerability implicit in the creative act to impinge upon him. Christians believe that one of the implications of the great symbol of the ascension of Christ is that in him humanity is taken into the divine nature. As part of that mystery, the Eternal accepts the experience of temporality.

Can such a stumbling discussion make any sense? Two things need to be said. In science when we discuss physical systems, we are concerned with objects that in some sense we transcend. Pascal said, "Man is only a reed, the weakest thing in nature; but he is a thinking reed."[17] That power of thought is what enables us to master and understand the physical world, putting it to the experimental test. In theology we are concerned with One who transcends us. There is a mystery in the nature of the Infinite which will never be comprehended by the finite. In theology there is a tradition called apophatic which acknowledges the essential ineffability of God, the One who is to be met only in clouds and thick darkness. This tradition has been stronger in the contemplative Eastern Orthodox Church than in the rationally confident Latin Church of the West. It is an essential corrective to all theological endeavors, though taken in isolation it would subvert all such endeavors. God is on the one hand unknowable, but on the other hand he has acted to make himself known. It is this latter fact that gives theology its mandate. Yet it will always be limited in its power to comprehend. Here is the explanation of the striking contrast in the relative successfulness of science and theology; the one advancing to greater understanding, the other continually wrestling with age-old problems.

The second thing to say is that the inevitable mystery in the nature of God is not a license for irrational assertion about him. Reason has its limitations, but it is not to be trifled with. Paradox may be forced upon us, our logic and our imaginations may be inadequate to complete the theological task, but we are only to embrace polarities which are required by experience. The insights of faith may not be demonstrable, or even wholly reconcilable, but they are not unmotivated. The tension between God's eternal nature and his involvement with the world, his timeless knowledge and acts of human choice, is closely allied to an important element in human religious experience. Writing to the Philippians, Paul exhorts them to "work out your own salvation with fear and trembling; for God is at work in you."[18] These words reflect a dialectical element in Christian experience: the paradoxical conviction both that we are responsible for what we do and also that it is God who is at work in our lives, molding and transforming them.[19] Before a decision it is the former that dominates our thinking; in retrospect it is the latter that we wish to acknowledge.

This confrontation of responsibility and grace is the human counterpart in our relation to God of the paradoxes we have been considering in his relation to us, his eternal purposes dependent for their fulfillment on the contingent response of men.

Theology and science differ greatly in the nature of the subject of their concern. Yet each is attempting to understand aspects of the way the world is. There are, therefore, important points of kinship between the two disciplines. They are not chalk and cheese, irrational assertion compared with reasonable investigation, as the caricature account would have it. The degree of their relationship is expressed by Carnes when he writes, "The activities of the theologian are as fallible and his theories as corrigible, as those of any other scientist and any other theories!"[20] He goes on to consider four "metatheological desiderata"; that is to say, qualities which should characterize the theological enterprise if it is to claim intellectual respectability. They are:[21]

(i) *Coherence.* The discourse must hang together. The ultimate achievement of this would be total consistency, but because of the considerations we have been discussing, theology may have to be content to live with some degree of paradox (just as science had to live for a while with the unresolved conflict between the wave and particle natures of light until it found the higher rationality of quantum field theory[22]).

(ii) *Economy.* Theology is not wantonly to multiply entities and explanations. This criterion might be thought to give preference to monotheism over polytheism.

(iii) *Adequacy.* Theology must be sufficiently rich in concepts to be able to discuss all its matters of concern.

(iv) *Existential relevance.* There must be an interpretative scheme which links theology with the actual content of religious experience.

Clearly there is a great deal here which is analogous to the demands made of a successful scientific theory.

The open character of religious experience, in which meaning is to be found only in the event itself and cannot be specified beforehand, implies that theological assertion will not be subject to simple verifiability (in the way that inductive arguments can be held to substantiate scientific statements) or to simple falsifiability (in the manner suggested by Popper as a scientific fall-back position for those unable to swallow the use of induction). Either of these positions in the philosophy of science requires the possibility of exercising predictive skills which are not available to the theologian. What then does the latter have to offer in defense of his subject's claim to be in touch with reality? In a famous essay John Wisdom suggests that theological statements are concerned to direct attention to patterns discernible in the facts:

It is possible to have before one's eyes all the items of a pattern and still to miss the pattern . . . And if we say as we did at the beginning that when a difference as to the existence of a God is not one as to future happenings then it is not experimental and therefore not as to facts, we must not forthwith assume that there is no right or wrong about it, nor rationality or irrationality, no appropriateness or inappropriateness, no procedure which tends to settle it, nor even that this procedure is in no sense a discovery of new facts. After all even in science this is not so.[23]

Reasons can be given for or against theological claims, just as they can be given for or against the theory of evolution in biology, though the latter also has not proved to have predictive power (and has been dismissed by Popper as nonscientific in consequence!). There is a close analogy with our claims to understand, at least partially, the personalities of other people. We do not have direct experience of their inner feelings, and however well we think we know what makes our friend tick, we shall not be able accurately to predict his behavior on every occasion. If he acts radically out of what we have believed to be his character, we shall either have to find some further explanatory circumstance (the kindly and responsible John's neglect of his mother was due to an overwhelming anxiety about his daughter) or revise our opinions (secretly John had always resented his mother). In judgments of personality—indeed, in all personal knowledge, including the knowledge of God—there are reasons for belief or disbelief, but because our knowledge is always partial, these reasons are never capable of being by themselves totally decisive. In the search for insight, we have to relinquish the claim to knock-down demonstration, though not to any possibility of a reasoned defense of the position we adopt.

Comparison with our knowledge of persons is apt. Carnes makes a valuable point when he says that theological statements are "cognitive but non-descriptive."[24] By that he means that they refer to a reality but can do so only in metaphorical terms. Our apophatic ignorance of the divine nature means that we can speak of God only by means of the language of analogy. There is an inescapably anthropomorphic element in our talk about God, not because we believe that he is an old man with a beard living above the sky, but because our experience of human personality touches the deepest levels of our being and so provides the least inadequate language to express our relationship with him who transcends personality.

There is one unusual claim for the verifiability of theological statements to which John Hick has drawn our attention.[25] He calls it eschatological verification. Christianity asserts that there is a destiny beyond death in which we shall come to know Christ and the Father with a clarity not given to us in this life. "For now we see in a mirror dimly, but then face to face. Now I know in part; then I shall understand fully, even as I have been fully understood."[26] Here is an unambiguous prediction of future illumination, but one that can be checked only postmortem. It does not assist us by providing a methodology for use in this world, but it does remind us that theological statements are not without claims to factual content of unusual kinds.

Just as there is a variety of assessments of the nature of science's achievement (pp. 24–28), so there is a variety of assessments of the nature of the achievement of theology. We are not concerned now with those who totally dismiss the reality or meaningfulness of religious experience, just as when we spoke of science, we were not concerned with those in some Eastern traditions who regard the whole physical world as illusion. It is possible for people to see that there is a component in man's experience properly labeled as religious but to disagree about what significance is to be associated with it.

Positivists see much scientific discourse, and in particular such rarefied subjects as quantum theory, as convenient manners of speaking whose real nature is simply to provide a harmonious correlation of observation. There are not really electrons or quarks. In a very similar way, there are people who see theological discourse as a manner of speaking about human needs—for example, the projection of an internal desire for reassurance in the face of a hostile world onto an external father figure. In this view theology is just disguised anthropology. There is not really a God. The modern founder figure of this point of view was Ludwig Feuerbach, who influenced many subsequent nineteenth-century thinkers. One was Karl Marx. He did not only describe religion, in that famous phrase, as "the opiate of the people," but also as "the sigh of the oppressed creature." For Marx, religion was the necessary articulation of the feelings of the exploited. It would wither away, just as the state would wither away, in a truly socialist society. Ironically, history, in which Marx placed his hope, has not vindicated his expectation. Another person influenced by Feuerbach was Freud. He saw religion as a universal neurosis or illusion, an expression of man's infantile need for comfort. His former colleague, Carl Gustav Jung, took a much more positive view. He discerned in the unconscious mind the activity of symbols which he called archetypes. They recur in many different ages and cultures. Jung saw them as part of the collective unconscious, the shared psychic heritage of mankind which is the suboceanic linkage of the ego islands of men. The greatest of these archetypes is that integrating symbol which Jung called the Self. It is hard to distinguish in Jung's thought between experience of the Self and experience of God.

Jung's ideas have proved helpful to many Christians in understanding their religious experience and conviction.[27] He himself, when interviewed in old age on television, on being asked by John

Freeman if he believed in God, said, "I don't need to believe, I know."
Yet it is also possible to see his ideas as providing the basis of an
anthropological takeover of God. His God is so often the God-image
in the psyche, not the One who stands over against us in judgment
and mercy. His thought is in many ways congenial to modern descen-
dants of Feuerbach, like Don Cupitt. Cupitt has a very high and passionate view of the importance of religion, but he seems to see it as a
purely human activity. "God simply *is* the ideal unity of all value, its
claim upon us and its creative power . . . To speak of God is to speak
about moral and spiritual goals we ought to be aiming at, and about
what we ought to become."[28] But if God is just a manner of speaking about the individual moral quest, it is difficult to see what is the
ground of the imperative we feel urging us to that quest. On the other
hand, if God is the creator of the world and through it is achieving
his purposes of love, then our perceptions of moral imperative will be
intuitions of his will and find their authority in that fact. When order
and intelligibility are introduced into any realm of experience by the
use of certain concepts, then that is prima facie a reason for believing
in the reality of the entities to which those concepts refer. Naturally
there are elements of projected human need in our talk about religion.
We have to sift our religious enquiry carefully for the contaminant of
wish-fulfillment. But to think that Feuerbach and his successors give
an adequate account of the knowledge and trust in God expressed by
Jesus and the saints is as implausible as supposing that quarks and
electrons are merely useful figments of the physicists' imaginations.

Instrumentalists see science simply as a means to an end. There
are also religious instrumentalists. An example would be the philosopher Richard Braithwaite who believed that the purpose of the Gospels and other religious stories was primarily to inculcate "agapeistic
attitudes,"[29] in other words to recommend a way of life conformed
to the ideal of Christian love. The intention of these stories, in his
view, is inspirational rather than informative. It no more matters to

Braithwaite whether they are true or not than it would matter to an instrumentalist whether a scientific theory which did the manipulative trick were true or not. There is a good deal of emphasis in contemporary theology on the idea of story.[30] We need to keep in mind that there is a significant difference between a story and a true story. The death of little Nell moves us in a different way from the dying of a child we know. The power of literature resides ultimately in its capacity to illuminate life, to speak to the way things are. Even a tale of a private world, like Tolkien's *Lord of the Rings,* derives its force from the reflection in an invented setting of the conflict of light and darkness more elusively present in the world of everyday. Scientific theories which consistently work are likely to do so because they represent with some degree of verisimilitude the structure of the physical world. Theological stories will only have power if they too mirror reality.

The scientific idealism of Kant and Eddington sought to settle issues by the exercise of pure thought. In a sense the counterpart in theology is provided by the belief that God's existence can be demonstrated by the use of reason alone. The "proofs" of God have a long history. The argument from design, beloved of eighteenth-century theologians, appeals to a rather detailed knowledge of the structure of the world. Other arguments, such as the need for a first cause (or its variant, the cosmological argument, which asserts that the very existence of the world demands a reason) are concerned with more general features. Today we do not feel that these proofs are very forceful. The intellectual buck has got to stop somewhere; why should it not do so with the origin of space and time and matter in the explosive singularity of the big bang, or with the assertion that the world just *is* and that is as good an unexplained starting point as that provided by postulating the existence of God? Far the most subtle of the purported demonstrations of the existence of God is the ontological argument, propounded by the greatest of the Archbishops of Canterbury, St. Anselm. Its merit is that it is purely logical, without appeal to any of the external features of the world. Anselm defines God as "something than which no greater can be conceived." We begin to list the characteristics of such a being: omnipotence, omniscience... Suppose we stop short of adding to that list "existence"? Now, is it not clear that something that exists is greater than something that does not exist? So we were wrong to have stopped just there if we are aiming for the greatest conceivable being. "Something than which no greater can be conceived" must have existence among its predicates. Hence God exists. QED. It is the sort of brilliant argument that has something fishy about it. Anselm has pulled a divine rabbit out of a logical top hat. Kant was the man who succeeded in giving the game away about how the trick was done. "Existing" is not really a predicate like "omnipotent." Its logical character is different. It simply asserts that there is an instance to be found of that entity which is characterized by the true predicates. But that, of course, is the point at issue, whether an actual instance of God is to be found. Anselm is not entitled to smuggle this in as if it were a defining predicate. Hence he has not achieved what he set out to do. God's existence remains a logically open question.

Almost all contemporary theologians would agree that the classical proofs of God's existence, regarded as incontrovertible demonstrations, do not work. That does not mean, however, that some of the issues to which those arguments direct us do not have considerable value as sources of possible insight into the way the world is. It is only their intellectual coerciveness which is to be rejected. After all, there is very little of interest, in either science or theology, which is capable of such absolute assertion. We shall have cause in later chapters to reexamine the cosmological argument and the argument from design in this more modest mode.

The idealist attitude in science is motivated by a desire to liberate scientific thought from bondage to detailed investigation of common experience. In a different manner theological thought which greatly emphasizes the concept of revelation seeks to free theology from involvement with common human experience. The outstanding proponent of such a theological stance in the twentieth century has been Karl Barth. His neo-orthodoxy proclaimed a divorce between revelation and reason. Such an attitude is in danger of turning theology into one of Wittgenstein's self-contained "language games." I believe that there is much material for theology to be found in the flux of everyday life and that our goal is an integrated picture of the way the world is. In that picture science and theology, reason and revelation, all find their place. There is indeed revelation of God, in those particular events and understandings preserved in scripture and tradition, but it is not insulated from the critique of reason or from evaluation in association with other forms of insight.

Theology differs from science in many respects because of its very different subject matter, a personal God who cannot be put to the test in the way that the impersonal physical world can be subjected to experimental enquiry. Yet science and theology have this in common, that each can be, and should be, defended as being investigations of what is, the search for increasing verisimilitude in our understanding of reality.

4

The Nature of
the Physical World

So far our talk has been in terms of generalities. It is time to become more specific. I shall attempt to sketch the picture of the world which science presents to us. This view will not extend in any inevitable way into a wider account of the way things are. Sometimes scientists, like Julian Huxley, have purported to be able to make such an extrapolation. It is clearly false. Counterexamples spring readily to mind. Jacques Monod[1] and Teilhard de Chardin[2] both survey the sweep of evolutionary biology. To one it is a tale told by an idiot, only capable of being confronted with a defiant nihilism. To the other its grandeur evokes a response of almost mystic intensity, as he peers forward to the climactic destiny of the omega point. The judgment between these conflicting reactions cannot be made on scientific grounds. In fact, there is no unique way of erecting a metaphysical structure on a physical basis, but equally such a basis is not capable of sustaining an arbitrary metaphysic. There has to be some degree of consonance between the scientific view of the world and those further considerations we may wish to add to it. That is why changes in our scientific understanding, such as the displacement of the earth from the center of the cosmos or the discovery that man has evolved from lower animals, have modified the tone of theological discourse in significant ways.

I want to suggest that the scientific view of the world that we currently hold is characterized by ten qualities. It is—

1. ELUSIVE. The world of classical physics, described by Newton and Maxwell, was clear and determinate. It was populated by particles of matter following definite trajectories and by aether waves in specific states of oscillation. It evolved in time according to strict laws. In this century that picture of the world has dissolved into the cloudy fitful world of quantum theory.[3] Its unpicturability is characterized by the duality of wave and particle so that its inhabitants share in natures which to the nineteenth-century physicist were totally incompatible, sometimes behaving like particles and sometimes behaving like waves. The outcome of an experiment is not usually completely determined; there are a variety of possible results. A probability can be given for the chance of obtaining any one of these specific answers, but no cause is to be assigned for its actual occurrence on a particular occasion. The solid dependable world of everyday becomes a shadowy unreliable world at its subatomic roots.

It is a critical question what reality is to be assigned to such a quantum world. Heisenberg allows an electron to have position (we can know where it is) or momentum (we can know what it is doing) but not both at once (we cannot visualize its state of motion as we can in classical physics). In what sense can such a protean object be called real? A variety of answers has been given. That grand old man of quantum theory, Niels Bohr, sometimes appeared uncompromising in his positivism. He was wary of making ontological pronouncements in public (of saying what *is*), but in conversation with a friend he once remarked: "There is no quantum world. There is only abstract quantum physical description. It is wrong to think that the task of physics is to find out how nature *is*. Physics concerns what we can say about nature."[4] In other words, it is all a manner of speaking. Personally I think that Bohr got that wrong. The researches of elementary particle physics have revealed an intricate and beautiful structure in the subatomic world, providing the foundation for our current belief that matter is composed of quarks and gluons.[5] It is difficult, to say the least, to think that all that is involved here is a particularly felicitous reconciliation of the behavior of electronic counters and tracks in bubble chambers. Any adequate account of what is going on must do justice to the impression of discovery which is so strong in the experience of the participants and provides them with the reward for all their weary labor. I believe that elementary particle physics discloses to us an actual reality. Yet it is certainly a reality of a more subtle kind than that corresponding to naive objectivity. Dr. Johnson kicking the stone to refute Bishop Berkeley will not do. That stone is almost all empty space, and what is not is a weaving of wave-mechanical patterns. Quantum theory represents a transformation of our understanding of the physical world. It is as though the program of Galileo and Locke, which discarded the secondary qualities of color, taste, etc., in favor of the primary quantities of dynamics, position, momentum, etc., has been carried a stage further, and these dynamical quantities are themselves to be thought of lying latent in a fundamental property which Heisenberg, borrowing a word from Aristotle, called *potentia*. An electron does not have simultaneous position and momentum, but its motion can always be described by a wave function which contains the potentiality for either.[6]

Heisenberg once wrote:

In the experiments about atomic events we have to do with things and facts, with phenomena which are just as real as the phenomena in daily life. But atoms or elementary particles are not as real; they form a world of potentialities or possibilities rather than one of things or facts.[7]

I question that he was right to think that this position reduced the claim of atoms and elementary particles to full reality. The grounds of my demur can be explained after we have considered the next property of the physical world. Science depends upon the fact that it is—

2. INTELLIGIBLE. This is so familiar that most of the time we take it for granted. Without it science would be impossible. In its most articulate form, it involves the use of mathematics as the basic expression of our understanding of the physical world. Something very odd is going on when this happens. Mathematics is the free discovery of the human mind. Our pure mathematical friends sit in their studies and think their abstract thoughts. They are constrained only by the requirements of consistency and certain canons of significance which assign value to general ideas capable of illuminating many particular mathematical instances. Yet some of the most intricate patterns they evolve prove to be just those realized in the physical structure of the world. It is an actual technique in fundamental physics to seek theories which are mathematically elegant and tightly-knit—in a word, which are mathematically beautiful—in the *expectation* that they will prove the ones to fit the facts. A theoretical physicist on being presented with a theory which is mathematically clumsy and contrived instinctively feels "That can't be right!" Paul Dirac wrote:

It is more important to have beauty in one's equations than to have them fit experiment . . . because the discrepancy may be due to minor features which are not properly taken into account and which will get cleared up with further developments of the theory . . . It seems that if one is working from the point of view of getting beauty in one's equations, and if one has a really sound instinct, one is on a sure line of success.[8]

In other words, it works. Mathematics is the abstract key which turns the lock of the physical universe.

It seems to me that it is a very significant fact about the world that the experienced rationality of our minds (of which mathematics is an expression) and the perceived rationality of the world (discerned by science) are consonant in this way. We have been concentrating on the explicit form of this relationship, revealed in developed physical theory. There is also an implicit form of the relationship, for it seems to be the case that those tacit acts of judgment involving skill (Dirac's

"really sound instinct") are also made in ways that actually are fruitful for our grasp of the structure of the world. There is this remarkable congruence between our inward thought and the outward way
things are. I have already given reasons for rejecting the view that the
order we find in nature is imposed upon it by us. Thus I find perverse
the claim by Andrew Pickering that "given their extensive training in
mathematical techniques, the preponderance of mathematics in particle physicists' accounts of reality is no more hard to explain than
the fondness of ethnic groups for their native language."[9] It is exactly
the other way round. Physicists laboriously master mathematical
techniques because experience has shown that they provide the best,
indeed the only, way to understand the physical world. We choose
that language because it is the one that is being "spoken" to us by the
cosmos.

The intelligibility of the world calls for an explanation. Einstein
said that the only incomprehensible thing about the world is that it
is comprehensible. The explanation will not be given us by science,
since science assumes the world's intelligibility as part of its initial act
of faith. In the next chapter we shall see what theology has to offer.
Meanwhile it is to intelligibility that I would wish to appeal in defense
of the reality of the quantum world. The theologian Eric Mascall
wrote:

The point is that though the physicist knows the objective world only
through the mediation of sensation, the essential character of the objective
world is not sensibility but intelligibility. Its objectivity is not manifested by
observers having the same sensory experience of it, but by their being able,
through their diverse sensory experiences, to acquire a common *understanding* of it.[10]

I think that is very much to the point. It is our ability to understand the physical world which convinces us of its reality, even when,
in the elusive world of quantum theory, that reality is not naively pic-
turable. This gives physics a good deal in common with theology as

the latter pursues its search for an understanding of the Unpicturable.

Despite the successes of science in comprehending the physical world we must also acknowledge that our view is—

3. PROBLEMATIC. The greatest paradox about quantum theory is that after more than eighty years of successful exploitation of its techniques, its interpretation still remains a matter of dispute.[11] We all agree how to do the sums, and our answers fit experiment like a glove, but we cannot all agree what is going on. I am not just referring to the contrast between the conventional indeterminate quantum theory and Bohm's determinate version (p. 14). Even if you accept my arguments for preferring the former, you are faced with conflicting views on what actually happens when we make a measurement on a quantum mechanical system.

Such systems are too small for us to have direct knowledge of them. An attempt, for example, to ascertain where an electron is depends for its success upon a chain of correlated consequence linking the position of the electron to some macroscopically observable signal, such as a mark on a photographic plate, which effectively says to us, "Here it is." This measurement requires, then, an interlocking of our dependable everyday world with the fitful world of quantum theory, in such a way as to produce, on a particular occasion, a particular result for the electron's location. Somewhere along the line from the quantum world to the ordinary world it gets fixed that we get that particular result on that particular occasion. The contentious question is, "How does this happen?" Four possible lines have been pursued in trying to frame an answer:

(i) One essentially declines to say. It regards quantum theory simply as a calculus of knowledge and does not worry about how that knowledge is obtained. To my mind this approach abdicates from the proper task of physics, to seek an understanding of the world.

(ii) Another is the Copenhagen interpretation, hammered out by Niels Bohr and his colleagues and subsequently prescribed as a rigid orthodoxy for the adherence of the faithful. This approach lays great stress on the role of measuring apparatus, those large instruments to be found in the everyday world of the laboratory which reliably register results without a quiver of quantum uncertainty. According to the Copenhagen interpretation, the arrangement of this measuring apparatus is always to be annexed to the description of every quantum mechanical investigation (this is the celebrated assertion of the indivisibility of observer (measuring apparatus) and observed (electron)) and in it they play the objectifying role of determining its outcome. Personally, I feel that the right interpretation is to be found along these lines, but that the Copenhagen School failed adequately to acknowledge how problematic it is to understand in detail the role of these classical measuring instruments. After all, they are composed of components (ultimately elementary particles) which themselves are subject to quantum mechanical fitfulness. There is a tension between the dependability of the instruments and the uncertainty of their constituents which makes it far from clear how things can be fitted together in this way with total consistency.

(iii) Because of this problem, others have continued the chain of correlated consequence until it reaches the point where a conscious observer intervenes to note the result. Here, surely, in consciousness (that interface between mind and matter) we perceive the coming into play of something new in the measuring process. May this not, then, be the point at which the question of a definite result gets settled? Certainly common sense requires that at the latest the issue must by then be fixed. The tale of Schrödinger's cat makes the point.

This unfortunate animal is incarcerated in a closed box which also contains a radioactive source with a 50/50 chance of decaying within the hour. If the decay takes place, it triggers the release of poison gas which instantly kills the cat. At the end of the hour, as an external observer we can say prior to opening the box only that we have an even chance of finding the animal alive. Yet we feel certain that the cat itself knows whether it is dead or not. The intervention of its consciousness must be the latest link in the chain at which the outcome of the radioactive decay becomes determined. We cannot suppose it all depends on us when we open the box.

At first sight it might seem attractive to assign this objectifying role to the first intervention of consciousness. However, it leads to some very counterintuitive conclusions. Are we to suppose that the computer printout recording the result of a quantum experiment and stored away for a while unread acquires a definite imprint only when someone opens the drawer and begins to read it? That is a pretty odd thing to believe.

(iv) Even odder is the fourth suggested interpretation proposed by Hugh Everett III. He suggests that at every act of quantum measurement capable of yielding a variety of outcomes, the world divides up into a series of parallel but disconnected worlds, in each of which one of these possible outcomes is realized. There is a world in which Schrödinger's cat lives and a world in which it dies. Because these subsequent worlds are disconnected, the two cats are unaware of their common past in the original feline. Since acts of quantum measurement are happening all the time, the resulting multiplicity of worlds is continually augmenting at a stupendous rate. Commenting elsewhere on this outrageous suggestion I wrote:

It is enough to make poor William of Occam turn in his grave. Entities are being multiplied with incredible profusion. Such prodigality makes little appeal to professional scientists, whose instincts are to seek for a tight and economical understanding of the world. Very few of them indeed have espoused the Everett interpretation. It has, however, been more popular with what one might call the "Gee-whizz" school of science popularizers, always out to stun the public with the weirdness of what they have to offer.[12]

If you thought that scientists always knew what they were up to, you may have been surprised by this variety of conflicting interpretations on offer concerning a subject as basic as quantum theory, and in its consequences as well established. It is part of a wider aspect of the scientific view of the world, namely that it is often—

4. SURPRISING. As new regimes open up, it is a regular experience to come across the new and totally unforeseen. At the end of the nineteenth century, William Thomson, Lord Kelvin, set out to calculate how long the sun had been shining. Kelvin understood classical physics as well as any man has ever done. To his mind there was only one conceivable source of the energy necessary to maintain the sun's output, namely that made available by gravitational contraction. As the sun shrank, the gravitational forces pulling it ever tighter together would do work which could keep the solar furnace burning. It was not difficult to calculate on that basis the maximum time that the process could have been kept going. Kelvin did his sums correctly and came up with the answer of about twenty million years. Obviously this contradicted the period of many hundreds of millions of years required by the students of the fossil record. A paleontologist had the temerity to ask the grand old man of physical science whether there could possibly be some hitherto unknown source of energy operating in the sun which could have prolonged its life. Secure, as he believed, in the certainties of nineteenth-century classical physics, Kelvin declared that to be impossible. We now know that it is the process of nuclear fusion, converting matter into energy, which has kept the sun shining for about five thousand million years.

It was no discredit to Kelvin that he did not know that. Rutherford had not yet discovered the nucleus nor had Einstein yet written down $E=mc^2$. However, Kelvin should have been more open to the possibility of processes beyond those then known to science. The world is full of surprises, and our powers of intellectual prevision are pretty myopic. The question which science typically asks is not, "What is it reasonable beforehand to suppose?" but rather, "What have we evidence to think is actually the case?" It is this continual encounter with the unexpected as new regimes come to be explored which gives to science its excitement and sense of discovery.

Next, we must acknowledge that the history of the physical world as we understand it is characterized by the unfolding of a process in which there is a continual interplay between—

5. CHANCE AND NECESSITY. It seems that a random event—a coming together of atoms in a particular pattern or the mutation of a gene—is the source of novelty in the evolution of complex systems. Such new possibilities would just vanish again were they not selectively perpetuated by competitive process acting in an environment of lawful necessity. Without chance there would be no change and development; without necessity there would be no preservation and selection. They are the yin and yang of evolution.

These ideas of chance and necessity were presented to the general public by Jacques Monod in his famous book of that title.[13] He pictures the evolution of life as arising from the chance aggregation of simple molecules into complexes capable of replicating themselves. Here is the action of chance. These replicating molecules rapidly reproduce themselves through the regularity of their chemical interaction with the environment, "gobbling up" the simple molecular food it provides. Here is the preserving role of necessity. A similar process is pictured in the neo-Darwinian account of evolutionary biology. Genetic mutations (chance) are selected by competition within a stable environment (necessity).

Two general questions can be asked about the process. The first is, "Is it adequate to explain the phenomena?" In particular, can a process of exploration by random shuffling (which is what is meant in this context by chance) be expected to produce in the available time scale such impressively complex results as those that we see around us and that we ourselves are? This problem is particularly acute in respect to the beginnings of life itself. It seems likely that there was a period of less than one thousand million years between the Earth's being sufficiently cool and in other ways suitable for the evolution of life and the actual coming into being of living material. That may seem a long time, but very complicated things have to happen. Fundamental constituents of all living beings are the proteins. Each protein is a very long chain of much simpler molecules, called amino acids. In a famous experiment Miller and Urey showed how these amino acids could be formed by the effect of electrical discharges and radiation upon a simple chemical environment made up of nonorganic molecules such as water, methane, carbon dioxide, etc., which would have been present in the young Earth's atmosphere. It is a much greater problem to understand how these amino acids got strung together to form the protein chains, particularly since most chains randomly assembled would not have a biochemically useful structure. Fred Hoyle[14] compares the chance of getting just one protein (and there are about two hundred thousand different proteins in our cells) to the chance of solving the Rubik cube blindfold. He calculates these odds of fifty million million million to one by analogy with pattern formation by beads randomly strung along a string. It is highly doubtful that this is the right way to think about protein formation. There may well be biochemical pathways, building chains from subchains, and subchains from sub-subchains, which make the process much more probable. We presently do not know what they are, and the distinguished molecular biologist, Francis Crick, was so little able to conceive what they might be that he thought it is impossible to understand how life could have evolved on Earth in the time available.[15] Instead, he believed (as does Hoyle in a very different way) that life arrived on Earth from elsewhere. It is not at all clear why what is inconceivable here was able to happen somewhere else, but so desperate a remedy as Crick's life-sent-in-a-space-capsule indicates the severity of the problem.

Not all scientists agree. For example, the theoretical biologist, Manfred Eigen, writes that:

The evolution of life, if it is based on a derivable physical principle, must be considered an *inevitable* process despite its indeterminate course . . . The models treated . . . and the experiments discussed earlier in the article indicate that it is not only inevitable "in principle" but also sufficiently probable within a realistic span of time.[16]

In other words, to Eigen the evolution of life is certain to take place even if in its precise details it depends on contingent circumstances. Some are so singularly confident that the process runs easily that they suppose the universe to be full of inhabited planets where life has taken forms, in detail different from what we know on Earth, but in general terms sufficiently similar for it to be sensible for us to search for signals emanating from some of the more advanced extraterrestrial civilizations.

When experts disagree in this way, we laymen will rightly conclude that the answer is not yet known. There are many things that we are ignorant about concerning the evolution of life, but there is no reason to suppose that science will not eventually find the answers.

Once life is here on Earth, neo-Darwinism is at hand to explain how genetic mutations, competitively selected, lead to the diversity and complexity of form that we find around us. The time scale for this development is about three thousand million years, though it is only in the Cambrian period, six hundred million years ago, that a stage is reached capable of depositing identifiable macrofossils. I think that a number of physical scientists feel an uneasiness about whether the neo-Darwinian accumulation of small improvements is the whole story of how such intricacy of structure came to be in the time available. The fossil record appears to have a stop-go character, with long intervals of stability interleaved with short periods of quick and largely unrecorded change. The rapid enlargement of the hominid brain size, with its eventual capacity for such not obviously selectively favored abilities as mathematical genius and artistic creativity, is not easy to understand. A. R. Wallace, who independently of Darwin formulated the principle of natural selection, wrote, "Natural selection could only have endowed savage man with a brain a little superior to an ape, whereas he actually possesses one little inferior to that of a philosopher."[17] To acknowledge these problems is by no means to hanker after a creationist notion of discontinuity caused by occasional supernatural intervention. It is simply to recognize (shades of Kelvin!) that there may be more to the story than the neo-Darwinists tell us. If so, it is the task of science to seek out what that extra might be.

The second question raised by the interplay of chance and necessity is the metaphysical one of what this might imply about our wider understanding of the world in which we live. Characteristically a variety of answers is on offer. Monod felt that the role of chance subverted all claims to significance in the processes of the world. He wrote, "Pure chance, absolutely free but blind, is at the very root of the stupendous edifice of evolution,"[18] and he concludes his book with intense Gallic rhetoric:

The ancient covenant is in pieces; man at last knows that he is alone in the unfeeling immensity of the universe, out of which he emerged by chance. Neither his destiny nor his duty have been written down. The kingdom above or the darkness below; it is for him to choose.[19]

Ah yes, but is there not something odd, or even marvelous, in the emergence of choosing beings through the playing of a cosmic game of roulette? When I read Monod's book, I was greatly excited by the scientific picture it presented. Instead of seeing the role of chance as an indication of the purposelessness and futility of the world, I was deeply moved by the thought of the astonishing fruitfulness that it revealed inherent in the laws of atomic physics. Those basic laws are just Maxwell's equations (to express the forces of electromagnetism controlling the larger scale structure of matter) and the Schrödinger equation (to express the quantum theory necessary for molecular dynamics). I could literally write them down on the back of an envelope. Yet the fact that they can have such remarkable consequences as you and me speaks of the amazing potentiality contained in their structure. From this point of view, the action of chance is to explore and realize that inherent fruitfulness. Arthur Peacocke wrote, "This role of chance is what one would expect of the universe were it so constituted as to be able to explore all potential forms of the organization of matter (both living and non-living) which it contains."[20] He goes on to speak in lyrical terms of the elaboration of life being like the development of a great fugue: "Thus does J. S. Bach create a complex and interlocking harmonious fusion of his seminal material, both through time and at any particular instant . . . In this kind of way might the Creator be imagined to unfold the potentialities of the universe which he himself has given it."[21] Such contrasting views of the world as those articulated by Monod and Peacocke have a long history which antedates the scientific insights which influenced these particular authors. In the nineteenth century Nietzsche wrote:

Perhaps there exist neither will nor purposes, and we have only imagined them. Those iron hands of necessity which shake the die-box of chance play their game for an infinite length of time. So there *have* to be throws which exactly resemble purposiveness and rationality of every degree.[22]

Curiously Nietzsche, the glorifier of the will, was speaking of human volition, but transposed to a cosmic scale his words would be acceptable to Monod, excluding only the assumption of an infinite time scale. On the other hand the fourteenth-century visionary, Julian of Norwich, wrote: "Indeed nothing happens by luck or by chance, but all is through the foresight and wisdom of God. If it seems luck or chance to us it is because we are blind and shortsighted."[23] These conflicting attitudes to the world, the denial or the affirmation of its significance, are certainly not just products of modern scientific ideas about chance and necessity. They have long been present in the thoughts of men. The verdict is not to be determined by science alone, a thought that is as true of those who deny significance as of those who affirm it. We shall return to these matters in the next chapter.

After such heady considerations our next property of the physical world may seem a trifle naive. The universe is awfully—

6. BIG. Our sun is an ordinary star among the hundred thousand million stars of our galaxy, which itself is not much to speak about among the hundred thousand million galaxies of the observable universe. The vastness of the world and the apparent insignificance of man is an idea which has been around for thousands of years. The psalmist wrote, "When I look at thy heavens, the work of thy fingers, the moon and the stars which thou hast established; what is man that thou art mindful of him, and the son of man that thou dost care for him?"[24]

Nevertheless, there is something specially chilling in the thought of the immensity of the world disclosed to twentieth-century science. Can there really be any significance attached to the lives of inhabitants of a speck of cosmic dust?

It would be a foolish error to confuse size and significance. John Baker wrote of such a tendency to be dismayed at the size of the cosmos and to assert that it destroys the possibility of a purpose at work in the world, "this small anthropocentric criticism is not an argument. It is nothing more than the complaining voice of mean, utilitarian, gutless, heartless, cerebral, twentieth-century, profit margin, Western man."[25] In science we are familiar with the fact that small effects are sometimes triggers for the discovery of large measures of new understanding. It was a tiny discrepancy in the orbit of the planet Mercury which first assured Einstein that his new theory of gravitation was to replace the work of Newton. Moreover, there is an interesting scientific insight into the size of the universe which links it directly with our presence within it. What that insight is can be explained only after we have considered the next property of the world. In its structure it is—

7. TIGHTLY KNIT. In recent years we have come to an understanding that if the cosmic process is to be capable of yielding systems as complex and as interesting as men, then delicate conditions have to be satisfied.

In the beginning was the big bang. As matter expanded from that initial singularity, it cooled. After about three minutes, the world was no longer hot enough to sustain universal nuclear interactions. At that moment its gross nuclear structure got fixed at its present proportion of three quarters hydrogen and one quarter helium.[26] Expansion and further cooling continued. Eventually gravity condensed matter into the first generation of galaxies and stars. In the interiors of these first stars, nuclear cookery started up again and produced heavy elements like carbon and iron, essential for life, which were scarcely present in the early stages of the universe's history. Some of these first generation stars exploded when they died, spewing out this new heavier matter into the environment. As second generation stars and planets condensed in their turn, on at least one of them there were now conditions of chemical composition and temperature and radiation permitting, through the interplay of chance and necessity, the coming into being of replicating molecules and life. Thus evolution began on the planet Earth. Eventually it led to you and me. We are all made of the ashes of dead stars.

For all this to be possible, there has to be a delicate balance of circumstances. The fundamental forces of nature have to be related to each other in just about the way that we actually find them to be. For example, the gross nuclear structure of the world, the proportion of hydrogen to helium, got fixed in the first three minutes of its evolution. This proportion depends for its exact value on the mass difference of proton and neutron and the ratio of the weak force (responsible for particle decays) to the other forces of nature. Change that ratio and the proportions of hydrogen and helium vary sensitively. If the weak force were weaker, and the decays in consequence went slower, then there would be all helium and no hydrogen. In such a world, without hydrogen, there would be none of the water which seems so essential to life. If the decays went faster, then for more complicated reasons supernovae explosions (distributing the first-generation heavy elements so as to be available in the environments of potentially life-evolving second-generation stars and planets) would probably be impossible. In that case the carbon and iron needed for life would be locked away useless in the cores of dying stars. Thus if life is to have a chance to get going, the magnitude of the weak decay force must be more or less what we see it to be.

Also, if life is to be possible, the world has to be about as big, and about as particular in its initial stages as it emerges from the big bang, as we actually find to be the case. The reason for the necessary size is that a world much smaller would have run its course before life had time to appear. It takes about fourteen thousand million years to make men, both because the evolution of complex life takes time and also because it can get going at all only in the second generation of stars and planets. This realization gives a surprising twist to our contemplation of the immensity of the universe. Without all those trillions and trillions of stars, we should not be here to be dismayed by them!

In the early expansion of the universe, there has to be a close balance between the expansive energy (driving things apart) and the force of gravity (pulling things together). If expansion dominated, then matter would fly apart too rapidly for condensation into galaxies and stars to take place. Nothing interesting could happen in so thinly spread a world. On the other hand, if gravity dominated, the world would collapse in on itself again before there was time for the processes of life to get going. For us to be possible requires a balance between the effects of expansion and contraction which at the earliest epoch science can speak of (the Planck time, 10^{-43} seconds) would differ from equality by not more than 1 in 10^{60}. The numerate will marvel at such a degree of accuracy. For the nonnumerate I will borrow an illustration from Paul Davies of what that accuracy means.[27] He points out that it is the same as aiming at a target an inch wide on the other side of the observable universe, fourteen thousand million light years away, and hitting the mark!

It is currently believed that the appropriate balance could have been induced shortly after the Planck epoch by a process proposed by Alan Guth, called inflation.[28] He suggests that it could be achieved naturally by a process called inflation. At a very early epoch, the fundamental laws of physics could have produced a phase change (crudely speaking like the change from water to steam) in which the universe was very quickly blown up very greatly in size. He claims that such a process would not only produce this near perfect balance of expansion and contraction, but also explain some other puzzling features of the cosmos, such as its isotropy. (It looks the same in all directions, even when we compare two diametrically opposed regions of the universe which are too far apart ever to have influenced each other.)

When someone tells you about processes he believes took place within incredibly tiny fractions of a second after the big bang (Guth deals with times less than 10^{-35} sec.), you should listen to him with respectful attention. He is a bright young man who is combining great ingenuity with a powerful grasp of the fundamental laws of physics. You should also listen to him with a certain wary reserve. He is describing to you the behavior of matter in regimes way beyond our direct experience. It is wise not to forget our limited powers of vision in these matters. We have had a great deal of difficulty in understanding after the event (let alone predicting beforehand) the behavior of matter in regimes closer to hand. The phenomenon of superconductivity was totally unexpected before its discovery by Kammerlingh Onnes, and it took over fifty years to understand it.

While inflation is a speculative idea, one must also say that it is attractive. If it is correct, however, it does not remove the special character of the world necessary to give life a chance. Inflation itself is possible only because of the character and balance of the fundamental laws of physics which, on that view, then produce the necessary balance of expansion and contraction.

Insights like these that we have been discussing seem to indicate that a world capable of producing systems of the complexity and fruitfulness of conscious beings has to be tightly knit in its structure. It is a comparatively recent realization by science that this is so. It has been given the name of the *anthropic principle*; that a world containing men is not just any old universe, "specified at random" so to speak, but it has to have a very particular character in its basic laws and circumstances. The anthropic principle is almost like an anti-Copernican revolution, not restoring the Earth to the center of the cosmos, but linking the nature of the universe with its potentiality for the evolution of men. What one might make of it is something we will discuss in the next chapter.

Meanwhile, having thought about the world's past, let us see what science has to say about its future. The predictions that it offers are characterized by asserting that ultimately the prospect is one of—

8. FUTILITY. Let us first consider our immediate neighborhood. Our sun is shining by burning up its hydrogen fuel to form helium in the process of nuclear fusion. Eventually the hydrogen will all be used up. The sun will then pass into the red giant phase of stellar evolution, expanding to the size of about the Earth's orbit. All life here will then be destroyed, burnt to a frazzle. What we fear might be brought about by a man-made nuclear catastrophe will inevitably come to pass eventually through a solar explosion. It is not an immediate worry. The sun has enough hydrogen to keep it in its present phase for about

another five thousand million years. But in the end, the Earth will become uninhabitable.

To be concerned about that is, perhaps, to take too cosmically parochial a view. After all, there may well be life elsewhere in the universe and anyway, long before that catastrophe is imminent, people may be technologically competent to colonize other planets round other stars. Let us lift our sights a little and enquire about the eventual fate of the universe itself.

Two scenarios seem possible. Which occurs depends upon the exact nature of that near balance between expansion and contraction in the world. We are not exactly sure which will win in the end. If expansion prevails, the galaxies will continue to fly apart. Within themselves, they will gradually condense, forming gigantic black holes. After almost inconceivable lengths of time, these black holes will decay, via a process described by Stephen Hawking. In this scenario the universe approaches an eventual heat death. If, on the other hand, contraction prevails, the present expansion will eventually reverse and the galaxies come flying back together again. What began with the big bang will end with the big crunch. The universe returns to its singular melting pot.

Either way, the long-term prospects for the physical world look bleak. The theologian John Macquarrie wrote, "Let me say frankly, however, that if it were shown that the universe is indeed headed for an all-enveloping death, then this might seem to constitute a state of affairs so negative that it might be held to falsify Christian faith and abolish Christian hope."[29] I do not find the scientific prognosis as dismaying as Macquarrie appears to do. There seems little doubt, though, that theology does face the challenge of the eventual futility of the physical world. Again, this is something to which we must return in the next chapter.

The final two qualities to consider relate to the total scientific description of the world. First of all it is—

9. COMPLETE (within the terms it sets for itself). The one God who is well and truly dead is the God of the Gaps. His job was to pop up as the explanation, so-called, of what otherwise could not be understood. The advance of scientific knowledge has given him a fading quality, so that he has become a sort of divine Cheshire Cat. Not that there are not many things which we do not understand. We have had occasion to note the unresolved puzzles about the origin of life. However, it no longer seems plausible that there are scientific no-go areas, in which questions can be posed scientifically to which only a God of the Gaps could provide an answer. Scientific questions demand scientific answers and they seem to get them. As the theoretical chemist and devout Christian, Charles Coulson, briskly said, "When we come to the scientifically unknown, our correct policy is not to rejoice because we have found God; it is to become better scientists."[30] The demise of the God of the Gaps should not be lamented, least of all by theologians. If God is God he is to be found everywhere, not just in the murkier corners of the world he has made.

The success of science in unraveling the workings of the physical world is the great cultural achievement of our century. The story it has to tell is to be taken with the utmost seriousness. Nevertheless, it is not the only story worth telling. The scientific worldview is also—

10. INCOMPLETE. Science trawls experience with a coarse-grained net. It restricts itself to certain kinds of enquiry, and in consequence much that is of the highest significance eludes it. Science surveys a lunar landscape—clear, orderly, lifeless. The real gap is between talking of complex metastable reproducing systems and talking of people. There is more to the world than physics can ever express.

One of the fundamental experiences of the scientific life is that of wonder at the beautiful structure of the world. It is the pay-off for all the weary hours of labor involved in the pursuit of research. Yet in the world described by science, where would that wonder find its lodging? Or our experiences of beauty? Of moral obligation? Of the presence of God? These seem to me to be quite as fundamental as anything we could measure in a laboratory. A worldview which does not take them adequately into account is woefully incomplete. I know, of course, that there are those who in principle regard these aspects of human experience as mere epiphenomenal froth on the surface of a purely physical substrate. In chapter 6 we shall have to consider such a point of view in detail. For the present let us just note that whatever we say in our studies none of us lives his life outside as if that were true.

Scientism is the mistaken attempt to exalt science into a complete philosophy. It will not work, and scientists have always been among the first to recognize that. Peter Medawar wrote:

There is no quicker way for a scientist to bring discredit upon himself and upon his profession than roundly to declare—particularly when no declaration of any kind is called for—that science knows or soon will know the answers to all questions worth asking, and that questions which do not admit a scientific answer are in some way nonquestions or "pseudoquestions" that only simpletons ask and only the gullible profess to be able to answer.[31]

I have tried to sketch the scientific view of the world. It would only be candid to end with a warning. A colleague a hundred years ago would have painted a very different picture. May it not seem all very different again in a hundred years' time? Indeed, does not the "pessimistic induction" (p. 22) make it virtually certain that that will be the case? It is because of the openness of science to the unexpected that I have to acknowledge that this might be so. I think, however, that most of the insights I have described will still find some place in whatever understanding will by then have been attained.

5

Points of Interaction

People sometimes say that science is concerned with questions of mechanism, with posing the question "How?" Theology is concerned with questions of purpose, with posing the question "Why?" The kettle is boiling on the gas stove. The scientist offers his explanation: the combustion of hydrocarbons generates heat which raises the temperature of the water until its vapor pressure equals atmospheric pressure and then it boils. The teleologist, who here must stand for the theologian, offers his explanation: I wish to make a cup of tea. The distinction is useful as far as it goes, indicating that different levels of meaning are involved, but it does not take us very far. The distinctions can become blurred. Cybernetic devices are capable of mimicking purpose through mechanism. The central heating switches on. We can say, because a strip of metal in the thermostat contracted as the temperature fell or, in order to maintain the temperature of the room within narrow limits. The latter is achieved by means of the former providing negative feedback for temperature control.

Within ourselves, our bodies are physical entities; our minds form intentions which we seek to execute. Their interrelationship is far more subtle than the two-fold story that Cartesian dualism tries to relate. Similarly there is not a simple dichotomy between science and theology. They interact upon each other in various ways.

At times, they have appeared to be in direct conflict, making rival assertions that contradicted each other. The most famous historical occasions are concerned with the church's adverse reactions to Galileo and Darwin. With hindsight we can see that theology was making unwarranted claims to pronounce upon questions which were both posable and answerable in purely scientific terms—the nature of motion and the structure of the solar system; the development of life and the physical origin of man. We can also see that by its eventual relinquishment of those claims theology was freed to reassert important insights of its own on matters concerning which science was powerless to utter—the faithfulness of God which finds its pale reflection in the regular laws of nature; the sustaining power of God maintaining the world in existence and achieving his purposes through its development.

There are contemporary points of interaction between science and theology which some perceive as areas of conflict. They include religious claims about miracles and about a human destiny beyond the disintegration of the body in death. We shall have to consider them in due course.

The second point of interaction between science and theology arises from the curious way in which modern science seems, almost irresistibly, to point beyond itself. In the last chapter I sketched a view of the world characterized by order, intelligibility, potentiality, and a tightly knit structure. Such a beautiful harmony evokes thoughts which verge on the religious. Einstein certainly thought so. He once said:

In every true searcher of Nature there is a kind of religious reverence; for he finds it impossible to imagine that he is first to have thought out the exceedingly delicate threads that connect his perceptions. The aspect of knowledge which has not yet been laid bare gives the investigator a feeling akin to that of a child who seeks to grasp the masterly way in which elders manipulate things.[1]

Natural theology, the search for God revealed in the works of his creation, has a long history. It played an important part in Thomas

Aquinas' theological scheme. Not surprisingly, it appealed to the Christian founding fathers of modern physical science. Galileo asserted, "Nor is God less excellently revealed in Nature's actions than in the sacred statements of the Bible."[2] While Newton, in the general *Scholium* to the *Principia,* was bold enough to claim that "to discourse of God does belong to Natural Philosophy." Today natural theology is not a popular pursuit among theologians. They have been made wary by the way that the argument from design, so beloved by Paley and other eighteenth-century apologists, shipwrecked on the Darwinian rocks. They are also mostly somewhat ignorant of modern science. They can even be faintly patronizing in their attitude, like Eduard Schillebeeckx who wrote, concerning the way in which our perceptions are formed and filtered by the interpretative models which we use (our spectacles behind the eyes), "It is with our thinking [sc. theologians' thinking] as it is with the physical sciences—at first sight a curious comparison for most of us, who are laymen in that discipline and so expect from it little insight into life's affairs!"[3] Today it is the scientists, rather than the theologians, who seem to be concerned with the exploration of natural theology. Paul Davies went so far as to write: "It may seem bizarre, but in my opinion science offers a surer path to God than religion . . . There is more to the world than meets the eye."[4] It is this feeling of "more than meets the eye" which characterizes the response of many scientists to the universe they investigate. It is a rum world in which we live, too beautifully structured, it seems, to be without an overall meaning. It is only fair to point out, however, that this is not a unanimous response. I have already spoken of Monod's feeling of the futility of the world. For honesty's sake let me also give a similar reaction by a distinguished physical scientist. Steven Weinberg has written, "The more the universe seems comprehensible, the more it also seems pointless," and he continues, with some bitterness, "The effort to understand the universe is one of the very few things that lifts human life above the level of farce, and gives it some of the grace of tragedy."[5] The claims of natural theology clearly call for some further consideration.

A third point of interaction between science and theology is provided by the mutual influence of their habits of thought. We have already noted that science and theology, whilst concerned with radically different kinds of subject matter, are not quite as distinct from each other in their procedures for seeking knowledge (epistemology) and in their problems concerning reality (ontology), as the popular caricature supposes. Each is corrigible, having to relate theory to experience, and each is essentially concerned with entities whose unpicturable reality is more subtle than that of naive objectivity. Moreover, theology has to use analogy as one of its principal theoretical tools, and in ages of broader culture than the one in which we live science has played an important part as an enlarger of the human imagination. Here too are matters which will require our further attention.

A fourth point of interaction, indeed of total absorption, would be provided by the assertion that all non-scientific levels of meaning are ultimately subverted by a thoroughgoing scientific reductionism. This is the claim that in the end there is "nothing but" scientifically discerned reality. That challenge will have to be met in the next chapter. For the meantime, let us allow the existence of different levels of discourse, each with its own authority, and seek to understand how they might interact with each other.

Possible Conflicts

Science has sometimes played a surgical role in relation to theology, amputating from it excrescences which are not true parts of the body theological. Today we can see that Darwin offered a healthy corrective to unjustified claims that the Bible had foreclosed the answers to purely scientific questions. The resultant gain was not merely scientific, but also theological. Thereby the early chapters of Genesis were released to serve their valid purpose as mythical statements about human beings' creaturely dependence and their experience of alienation from God, the true ground of their being. In this symbolic form these ancient stories retain remarkable power to speak to us.

Science has sometimes played an antiseptic role, by its patient investigation undercutting pretentious claims to esoteric knowledge. St. Augustine tells us that he first became disenchanted with Manichaeism, the strange eclectic religion he had embraced as a young man, when he compared its attempts to explain the incidence of eclipses with the much more accurate predictions of the astronomers. He tells us: "I had read a great many scientific books which were still alive in my memory. When I compared them with the tedious tales of the Manichees, it seemed to me that of the two, the theories of the scientists were more likely to be true."[6] Thus it was that he was encouraged to embark on a further spiritual quest which led, some years later, to his conversion to Christianity.

I can write quite happily about that incident, but what about those scientists who believe that their ideas have equally cast doubt on the "tedious tales of the Christians"? There are a number of issues on which it might be thought that scientific knowledge has discredited Christian doctrine. We must now turn our attention to them.

1. ORIGINS. The claims that the universe originated in the big bang and that God made the world are clearly not in direct conflict, since they are different categories of statement. They would be directly comparable only if God were a cause in the same sense as the originating singularity of the big bang operates as a cause of all that follows it. Then it would truly be God *or* the big bang, in the way that fifty years ago cosmologists adjudicated between the big bang and steady state theories of the universe. Theology has always insisted that God is not a cause among causes or an object among objects. His relation to the world is wholly different from that of any of its participants. God is properly to be understood as the ordainer and guarantor of natural law. His role is to sustain the world in being. Creation

is properly understood as a continuing act of God's will which maintains the cosmos moment by moment. It is not just about some initiating instant.

Thus for me and many other scientists of Christian conviction it is God *and* the big bang; the why and the how in terms of the simple analogy with which I started this chapter. To think that God is somehow deprived of his Creatorly role by increasing completeness in the scientific account of the universe is simply to demolish the straw deity of the God of the Gaps. Equally no scientific development of itself can confirm that Creatorly role. It was unfortunate, therefore, that Pope Pius XII was moved in an address to the Pontifical Academy of Science in 1951 to speak as if the big bang theory of cosmology, with its dateable origin of the world, had some superior value for theology over other possible accounts of the universe's history. Such a Papal endorsement of a particular scientific theory was as embarrassingly misconceived as was the opposition by his predecessor Urban VIII to the ideas of Copernicus and Galileo.

It is also the case that no great theological stakes are involved in the suggestion about the origin of the universe which has been made by the physicist Alan Guth. He proposed what seems at first sight to be a sort of scientific version of creation ex nihilo. The conjuring trick of appearing to get something (indeed everything) for nothing was described by Guth as the universe's being a "free lunch." The way it might work depends upon the curious properties of the vacuum in modern physics. Classically the vacuum is just emptiness, nothing there, nothing happening. Heisenberg does not allow a quantum vacuum to be so inert. Each possible state of matter—photons, electrons, each different sort of quark, and so on—is described by a quantum field. The state in which all of these fields have their lowest energy is the vacuum, rock bottom. There are then no photons, electrons, quarks, etc., present, but that does not mean that nothing is going on. Quite the contrary, for the vacuum in quantum theory is a humming hive of activity. There are no permanent particles present, but there are continual fluctuations in which particles transiently appear and disappear. This is necessary because of a phenomenon called "zero point motion." It is most easily understood by thinking of a very elementary physical system, the simple pendulum. In classical physics its lowest state of energy is when the bob of the pendulum is at the bottom and at rest. Then we know where it is (at the bottom) and what it is doing (it is at rest). Heisenberg will not tolerate such exact knowledge. He demands, therefore, that the lowest energy state of the quantum pendulum involves a slight quivering—near the bottom and nearly at rest, but never precisely so. This quivering is the zero point motion. Augmented to the complexity of a quantum field, it produces the fluctuating vacuum that I have described.

Guth's idea capitalizes on this. Most of the fluctuations are small, but just very very occasionally there might be a large one. Before it has time to disappear again, Guth's inflationary process takes over and blows it up into the universe in which we live. Hey presto! Everything has come from nothing.

The idea is clever. It has all sorts of physical difficulties, and it is by no means clear that it is correct and could work. However, let us generously assume that this is how our world came to be. There are two comments to make. One is that only by an extreme abuse of language can a quantum mechanical vacuum be called nothing. Its nature is specified by the number of different quantum fields which are stated to be present (even if each is in its lowest energy state) and the laws of physics which they obey. In its fluctuating nature, it is more like a plenum than a vacuum. The second comment is that, while such an idea might well eliminate a deistic God whose only role had been to light the blue touch paper of the big bang and then retire, it in no way threatens the Creator of Christian theology. His sustaining role is inviolate. The laws of physics obeyed by the quantum fields, which have to be assumed before one can begin at all to talk scientifically about the process, are expressions of his will and purpose. He is the ground of physical process, not a participant in it.

Other questions of origin can also be sources of possible conflict. We have already surveyed the role of chance "shuffling"[7] as the source of novelty in the evolution of life, interacting with the lawful necessity required to select and preserve favorable variations. For people like Monod, this totally subverts any claim to a purpose at work in the world, since what emerges is not foreseeable in advance. Certainly it seems that there is a contingent character to the way life has come to be. The astonishing potentiality of the laws of atomic physics may favor the development of systems of great complexity, but the details of their form appear to depend on random circumstances. These include not only chance genetic mutations but also the intervention of unforeseeable external events, such as the collision of a gigantic meteor with the Earth which some believe changed the course of evolution by extinguishing the dinosaurs.

Of course, it is possible to deny that chance is really at work at all. To someone like Donald Mackay,[8] the word is just a shorthand for unknown causation. He believes that all is the unrolling of a plan under total divine control. Such a contention is logically invincible but not wholly persuasive. Why has God chosen to hide his hand under the appearance of randomness? More attractive is the attitude of Peacocke,[9] who affirms as part of God's plan a positive exploratory role for chance in the realization of potentiality.

The matter is further complicated by the fact that the relationship between chance and determinism, chaos and order, is surprisingly subtle. For example, it is quite possible for random events to lead to inescapable conclusions. The path is subject to chance; its end is not. As a very simple example of this happening, consider the following game, based on a simplification of an example suggested by Bartholomew.[10] Start by six throws of a die. In this way one randomly obtains six numbers, each of which lies between 1 and 6. Call the number of instances in which the number obtained is less than or equal to 3, the current score. It is obviously a number lying between 0 and 6. Repeat with a further six throws of the die, and define a new current score by the number of instances obtained which are less than or equal to the previous current score. Repeat, etc. The process is generated throughout by a random process, namely the throw of a die. Nevertheless, either trial or reasoning will show you the inescapable fact that eventually, sooner or later, the current score homes in to either 0 or 6, after which it stably repeats itself ad infinitum. To give the matter a further twist, remember that the apparently random falls of the die are themselves totally determined, but by dynamical details of the shaking which are too fine for us to be aware of. It is only because of this that we attribute the outcome to chance. There is an intricate and subtle interlacing of the levels of chance and necessity in the functioning of the world.

Theology has always been in danger of a double bind in relation to physical causation. A tightly deterministic universe, evolving along predetermined lines, seems to leave little room for freedom and responsibility. It is congenial only to a deistic indifference or to the iron grip of Calvinist predestination. On the other hand, too loose a structure dissolves significance. Meaning can drown in the rising waters of chaos. A world capable of sustaining freedom and order requires an equilibrium between these rigidifying and dissolving tendencies. The actual balance between chance and necessity, contingency and potentiality, which we perceive seems to me to be consistent with the will of a patient and subtle Creator, content to achieve his purposes through the unfolding of process and accepting thereby a measure of the vulnerability and precariousness which always characterize the gift of freedom by love.

A final question of origin relates to the nature of humanity. We accept that our ancestry involves a continuous chain through the pre-hominids back to simpler life and eventually to the emergent replicating molecules in the amino-acid soup of early Earth. Yet Christianity also claims a unique role for human beings who are able to be in communion with their Creator. Are these statements compatible? I think so. After all, humanity by its acquisition of self-consciousness has transcended his origins. Pascal said, "All bodies, the firmament, the stars, the earth and its kingdoms are not worth the least of minds, for it knows them all and itself too, while bodies know nothing."[11] Self-consciousness is a mysterious property of humankind. We all experience it even if we do not understand its origin. If the organization of matter can reach such a stage that it gives rise to the capacity to attain knowledge of itself, there is no particular difficulty in supposing it also capable of attaining knowledge of God. This point of view relies on acknowledging the psychosomatic unity of human persons, over against the Greek dualistic view which saw humans as essentially spiritual beings, temporarily trapped in the prison of physical flesh. Though such Greek notions have often influenced Christian theology, at its truest the latter has held to the Hebrew idea of the essential unity of the human person regarded, in often-quoted words of Wheeler-Robinson, as "an animated body not an incarnated soul." On this view there is no question of the addition of a soul as an extra spiritual ingredient at some stage of human evolution. Our evolved capacity to respond to our Creator is the highest and most striking illustration of that potentiality with which the physical world has been endowed.

2. GOD'S INTERACTION WITH THE WORLD. The God of Christianity is not remote. His people turn to him in prayer. Doubtless a substantial part of prayer is a meditative waiting upon God, but it also has its petitionary aspect. Not all of that is to be dismissed as a childish hankering after a celestial Father Christmas. Yet, in a scientific age can we really believe that God interacts with our lawful world?

It does not seem difficult to suppose him involved with persons, a persuading, sustaining, transforming presence in the depths of our being. The power of symbol which depth psychology recognizes to be within the unconscious mind can well include such divine influence. I would not wish to identify God with the God-image in the psyche, but that may well be a point of contact between him and us. Most Christians (and many others) do not find it difficult to pray for strength for themselves or others in need. Our psychic life is mysterious, and comparatively little of its mode of operation is understood. Nevertheless, the belief that God can meet with men and women and respond to them is consonant with our interior experience.

It is one thing to pray for strength in time of personal need. It is something entirely different to pray for rain or for the healing of an inoperable cancer pronounced terminal by the doctors. If we dare to make such prayers, they are likely to contain the clause "if it be thy will." Partly this is a necessary recognition that what we want may not be what we ought to be seeking. There are clashes of interest; the farmer's desire for rain opposes the vicar's hope for a dry day for the church fete. There are refusals of destiny; it may be that the acceptance of the imminence of death is the spiritual path to be trodden by the patient with the malignancy. But partly it is also that we in this modern age find it hard to picture God in detailed control of things in the way that was natural for our scientifically ignorant ancestors. We want to give him an escape clause, covering up thereby our own lack of expectation. For us rain is a byproduct of the operation of the great heat engine of the Earth's seas and atmosphere, not the opening of the heavenly water sluices. Disease is biochemical malfunction, not the chastening act of a watchful Creator. Is then God just a *deus absconditus*, absent from his world except for his interventions within the psyches of human beings?

Christian theology understands God's act of creation as being a free act of love. Such an act may well involve the acceptance by its author of some measure of limitation, a kenosis (emptying) as the theologians say, of divine power. After all, God's omnipotence can mean only that he is able to do what he wills *consistent with his nature*. Love characteristically tempers power of command. Both the lawful necessity of the world and the role that contingent chance has to play within it are aspects of that great creative act. Thus in accepting them, God may have, in his freedom, accepted a self-limitation which circumscribes his mode of action. Nevertheless, there seem to be three ways in which we can conceive of his influencing the world other than through the minds of men (and any other self-conscious beings there may be).

The first lies in exploiting the room for maneuver he has left himself in the laxity that quantum theory produces at the subatomic roots of the world. Individual quantum events are "uncaused"; it is only the statistical distribution of many such similar events which is prescribed by physical theory. Yet can we not suppose that if no sparrow falls to the ground without our Father's will, then equally no atom decays without it either? Will not God's power to act as the cause of uncaused quantum events (always cleverly respecting the statistical regularities which are reflections of his faithfulness) give him the chance to play a manipulative role in a scientifically regular world? Admittedly these events are microscopic occurrences, but they can act as triggers of macroscopic consequence (as when a nuclear decay kills Schrödinger's cat). Such a view has been vigorously propounded by the physicist-theologian, William Pollard.[12]

There might be some sort of truth here, but I am uneasy about it as the principal account of God's action in the world. There is an air of contrivance about the whole idea. It is interesting to note that similar suggestions have been made to explain how the physical phenomena of our brains might be reconciled with our mental experience of choice and responsibility. It is proposed that Heisenberg uncertainty might give us room to manipulate ourselves. Quite apart from arguments about whether this is consistent with the neurophysiological characteristics of neuron excitement (are synaptic firings capable of being effectively quantum phenomena?), it does not seem to me that these Houdini-like wrigglings of mind within the fairly tight straitjacket of physical phenomena have the right feel to be explanations of what is going on. We are back with a struggling ghost inside a now somewhat rickety machine. Both our relationship with our physical bodies and God's relationship with his physical world must, it seems to me, be subtler than that.

We use the notion of chance in at least three distinct ways. One is the radically uncaused phenomena of quantum physics which we have been discussing. This is chance at its scientifically irreducible. The second sense is when small hidden effects have disproportionately large visible consequences—the fall in one direction or another of a top unstably balanced on its point is said to be "by chance" because though we believe it has causes, we are unable to discern them. The third sense is when two unrelated causal chains impinge upon each other. There are good reasons why the wind blew the slate off the roof. I had my reasons for passing the house. It is "by chance" that these two events coincide in time and lead to my injury. This latter sense of chance is often called accident. Some believe that God acts in the world by a combination of foresight and ingenious prior fixing, bringing about such coincidences to produce occasions of significance. It is possible to think about some of the miracles in the Gospels, particularly the nature miracles, in this way. We read how Jesus spoke a word and the storm on the lake was stilled.[13] The energy in that Palestinian weather system exceeded that of an atomic bomb. We need not, however, assume that Jesus altered its meteorological structure in a direct act of enormous power. Rather, there was a divinely ordained significance in the coming together of the lapsing of the storm through natural process and Jesus' saying "Peace. Be still!" The disciples were right indeed to say "Who then is this?" even if we cannot go on with them to say further "even wind and sea obey him." That is how the account would run.

Accident is scarcely the word to use in such circumstances. Jung would have called it synchronicity. He believed that there was this power at work in the world bringing about significant coincidence within the regular physical framework. Interestingly, the distinguished theoretical physicist Wolfgang Pauli agreed with him.

My own reactions are ambivalent. I certainly acknowledge the fact of experience that there are remarkable coincidences which carry the deepest significance for those involved. It would be ungrateful to refuse to accept insights that come to us in that way. However, I find it difficult to understand how this comes to be, how God or synchronicity can be so effective in shaping the flux of physical events. That is not to say that it cannot happen but simply that one mystery has been replaced by another. It is characteristic of this whole discussion, I am afraid, that it is tentative and unsatisfactory. I regret that, but I am unable to remedy it. However, I am not totally downcast, for I am equally unable to explain how my mental experience (the very raw material of all my experience, when I come to think of it) is related to the physical structure of my body and in particular to the neurophysiological functioning of my brain. The action of drugs, or a smart tap to the head, establish that they are connected. In the next chapter we shall have to struggle with this problem. You will not be surprised to learn that I shall not be able to solve it, though I think there are some moderately helpful things one can try to say. If we do not understand how we ourselves interact with the physical world (though we feel certain that we do, and none of us treats himself as if he were a robot), then it is scarcely surprising that we do not understand how God might do so either. Indeed, it is quite possible that our mode of interaction, whatever that may be, bears some faint analogical relation to God's, though to push that analogy too far would lead to pantheism (the identification of the world with God) or panentheism (the notion that the world is in God though God exceeds the world). Neither seems to me to be satisfactory theologically, for they fail to do justice to the experience of the Otherness of God which is a basic religious insight.

The third possible way in which God might influence the world is that of direct intervention, the specific exercise of his will to achieve a particular end. This is a topic of sufficient importance to stand on its own. We need to consider the possible point of conflict provided by religion's assertion of—

3. MIRACLE. As far as Christianity is concerned, two things need to be said. The first is that the Christian is not committed to believing in the literal truth of every miraculous event recorded in the Bible. An understanding of the role of myth and legend enables us to accept some stories as just that, pictorially valuable but not historically accurate. I do not believe that the sun stood still for Joshua when Israel fought the Amorites,[14] though if a historical reminiscence lies behind this story it perhaps refers to an eclipse coinciding with the day of battle, in itself a remarkable piece of synchronicity. The most implausible New Testament miracle to my mind is the story of the coin in the fish's mouth.[15] Peter tells Jesus that they have to pay the yearly half-shekel tax, due from each Jew for the upkeep of the Temple. Jesus tells him to catch a fish. In its mouth will be a shekel coin which will pay the score for both of them. What makes that story so incredible is that it seems to show Jesus acting as a wonder-worker operating for his own advantage. That does not at all fit the picture of him that the Gospels give us. Perhaps the story is a joke which got into the tradition by mistake as if it were an actual happening.

That point being made, the second is that, nevertheless, Christianity cannot escape the challenge of the miraculous, for at its heart lies the assertion of the resurrection, that God raised Jesus from the dead. Here seems a clear point of conflict, since in our experience dead men stay dead.

The distinguished nineteenth-century physicist, Sir George Stokes, said robustly to those who question the miraculous: "Admit the existence of a God, of a personal God, and the possibility of the miraculous follows at once. If the laws of nature are carried on in accordance with his will, he who willed them may will their suspension."[16] That is all very logical as far as it goes, but it does not go far enough. What it fails to take account of is that there is a *theological* problem about miracle. The question is not, "How can they happen?" but, "Why do they not happen more often?" God frequently seems just to let things take their disastrous course. If he intervenes he seems to do so sparingly, only occasionally obliging with direct help. Such a view, if true, would destroy the credibility of the God of steadfast love proclaimed by Christian theology. Indeed. intervention is not a word that one can properly use of such a God. Whatever his relation to his world, it must surely be faithful not capricious, regular rather than intermittent. While miracles are logically conceivable as divine tours de force, they are theologically inconceivable on that basis. They can find theological acceptance only if they are perceived as part of a wider unity of divine action and purpose, which goes beyond the experience of everyday but which forms with it a coherent whole. The pursuit of such a unified understanding is the main task of theology in the face of the unexpected which we call the miraculous.

This is not a pursuit wholly unfamiliar to the scientist. Science often faces the unexpected. Under normal circumstance the conduction of electricity in metals is adequately described by Ohm's law. The current flowing is proportional to the applied voltage, which is needed to overcome the resistance of the metal. Without such an applied voltage, no current flows at all. It was a great surprise when in 1911 Kammerlingh Onnes, having cooled down metals to hitherto unattained low temperatures, found that, for some metals below a certain critical temperature, they lost all their electrical resistance. A current would flow for many hours without any voltage being needed to sustain it. This discovery of the superconducting state did not mean that physics had suddenly become irrational. It was simply the case that there was a higher rationality than that known in the everyday world of Ohm. After more than fifty years of theoretical effort, an understanding of current flow in metals was found which subsumed both ordinary conduction and superconductivity into a single theory. The different behaviors correspond to different regimes, characterized by different organizations of the states of motion of the electrons in the metal. One regime changes into the other by a phase change (as the physicists call it) at the critical temperature.

The notion of regimes is important. Physics certainly does not say that all things always behave in the same way. Rather it says that similar circumstances lead to similar consequences. When we change from one regime to another (from conduction to superconductivity) radically different results can follow (the resistance vanishes). Changing circumstance can lead to totally unexpected effects.

Christianity claims that God was in Christ in a unique way. If that is true, it is to be expected that unprecedented events might occur, for Jesus represented the presence of a new regime in the world. Along these lines I believe that it is possible to form a coherent picture of God's activity in the world that embraces both the fact that in our experience dead men stay dead and also that God raised Jesus on Easter Day (and, I believe, that the tomb was found empty). It is not my purpose here to pursue that matter further.[17] It is theologically obligatory to seek such a coherent understanding of any other miraculous events which are claimed to be true against common expectation. If theology is successful in exhibiting the specific features which would justify the claim that in such events special regimes are present, and also in explaining why in these and *all other similar circumstances* God can consistently be supposed to act in the way asserted, then it has done all that it needs to do. Science is not in a position to contradict these special cases on the basis of its generalized investigations. On this view, as in the Fourth Gospel, miracles are seen, not as celestial conjuring tricks, but as signs, insights into a deeper rationality than that normally perceptible by us.

Christians believe that the resurrection of Jesus was out of the way of common experience by way of being an anticipation of what is to be the experience of all men. We "look for the resurrection of the dead."

4. FUTURE LIFE. Is the hope of a future life another point of conflict with science? It is important before attempting an answer to be sure about exactly what hope we are talking about. It has frequently been spoken of in terms of the immortality of the soul. Yet the characteristic New Testament hope is not for the continuance after death of a tenuous spiritual component of our being, but the restoration by God of the whole person. God "raised the Lord and will also raise us up by his power."[18] We have already had cause to note (p. 83) that the preferred Christian attitude to the human nature is the Hebraic acceptance of unity rather than the Greek divorce between soul and body. For those who accept the psychosomatic integrity of human beings, the future hope is bound to be resurrection rather than survival, the reconstitution of the whole person in some other environment of God's choosing.

Clearly such an idea goes beyond our direct experience, but it seems to me in no way to run counter to it. There is nothing particularly important in the actual physical constituents of our bodies. After a few years of nutrition and wear and tear, the atoms that make us up have nearly all been replaced by equivalent successors. It is the pattern that they form which constitutes the physical expression of our continuing personality. There seems no difficulty in conceiving of that pattern, dissolved at death, being recreated in another environment in an act of resurrection. In terms of a very crude analogy, it would be like transforming the software of a computer program (the "pattern" of our personality) from one piece of hardware (our body in this world) to another (our body in the world to come). Scientifically, this seems a coherent idea. Paul Davies, who cannot readily be suspected of being unduly influenced by Christian theology, has written, "Though some of these ideas [sc. the relation of information-bearing pattern to mind] may seem fearsome they do hold out the hope that we can make scientific sense of immortality, for they emphasize that the essential ingredient of mind is *information.*"[19] I think it is fair to claim, in terms of the traditional theological language I am using, that it is the resurrection possibility of reembodying mind that Davies is referring to when he speaks of immortality.

Natural Theology

The traditional role of natural theology was to provide demonstrations of certain basic theological truths, such as the existence of God. It was held that no reasonable human could fail to assent to its propositions. Further insight into the divine nature could then be added from the truths furnished by revelation.

Just as the view that science is made up of ineluctable theoretical prediction confronting unambiguous experimental fact has proved too simplistic and unsubtle an account, so this sharp division in theology between reason and revelation has been found unsatisfactory. There is certainly an important role for revelation, understood as the historic record of experiences of particular illumination, rather than a guaranteed packet of timeless propositions. God is everywhere at work, but his presence and activity are more clearly to be discerned in some circumstances than in others—just as well-designed experiments exhibit clearly scientific laws which operate in more covert form in less carefully selected situations. Above all, the Christian doctrine of the incarnation speaks of God's self-disclosure in the person of Jesus Christ. Yet every act of revelation takes place in a cultural setting, and its expression is formed from the particular point of perspective that imposes. It is part of the doctrine of the incarnation that Jesus was truly man, accepting thereby the limitations inherent in being a Jew in first-century Palestine. The New Testament, however remarkable the experience and understanding it preserves, is a document of its time. In every sacred text the gold of eternal truth is mixed with the base matter of contemporary attitudes. In making its truths their own, every generation has to do so by the exercise of reason in conjunction with the historically mediated data of revelation.

If theology is not to be compartmentalized into reason and revelation, nor is the world to be divided between the sacred and the profane. They interpenetrate each other. We are concerned not with nature and supernature, but with one reality. The coherence and integrity of God's relationship with the world, which we sought when we discussed miracle, is to be the basis of the whole theological endeavor.

Natural theology is important, because if God is the Creator of the world, he has surely not left it wholly without marks of his character, however veiled. There must be a consonance between the assertions of science and theology about the world in which we live. In seeking that consonance, we must certainly use reason, but we must also be prepared to recognize that in theology, as in science, reason will need to have conjoined to it tacit acts of judgment which fall short of proof beyond a peradventure. Natural theology cannot claim a universal certainty; like other branches of theology it partakes of the nature of faith. Its mode is insightful rather than demonstrative. The understandings it offers are neither irresistibly compelling nor unmotivated. The criticisms of Hume and Kant and Darwin disposed of the logically coercive claims of the classical "proofs" of the existence of God. They did not thereby remove their potential value as insights into a world in which there is much to suggest the presence of "more than meets the eye."

The cosmological argument was grounded on the assertion that the very existence of the world called for an explanation and that that explanation could come only from the world's being the creation of a God who possessed being itself (and therefore stood in no need himself of further explanation). It was not wholly clear why this irreducible character of unexplained being should not be attributed to the material of the cosmos itself. As a knock-down demonstration, the cosmological argument fails to convince. The cosmological *insight*, on the other hand, focuses on the intelligibility of the world, the way in which its patterns interlock with the patterns of our thinking, so that we can understand the world in which we live. The rational order that science discerns is so beautiful and striking that it is natural to ask why it should be so. It could find an explanation only in a cause itself essentially rational. This would be provided by the Reason of the Creator, which establishes the common ground for the observed rationality of the world and the experienced rationality of our minds, guaranteeing their mutual coherence. This insight gains cumulative force if we widen our view beyond that of science to recognize that we know the world also to contain beauty, moral obligation, and religious experience. These also find their ground in the Creator—in his joy, his will, and his presence.

The argument from design appealed to the intricate adaptation of animals and humanity to life in their environment. This was seen as compelling evidence that they were the planned products of a Creator's will. Darwin undercut that completely when he showed how natural selection could lead to "design" without a Designer. The *insight* of design looks beyond evolution to the astonishing potentiality of the simple laws of matter, capable of being explored by the processes of chance and necessity, and to the tightly knit character of a universe satisfying the anthropic principle. Again, it is natural to ask why the world should be so special. Scientists have felt particularly uneasy about the delicate balances required by the anthropic principle. To alleviate their anxiety, some of them have suggested that there might be a portfolio of many different universes. Either these are just independent worlds, "side by side" so to speak, or they can be conceived of as arising from an infinite series of oscillations of one universe, ever expanding and contracting and each time having its basic structure dissolved in the melting pot of the big crunch, thence reemerging in a different form in the subsequent expansion of the next big bang. Successive oscillations would in this way produce successively different worlds. If there are many different worlds on offer, with different laws and circumstances, then it would hardly be surprising if one of them had by chance the right balance to satisfy the anthropic principle. That is the one in which we live because we could not be around in any other. Let us recognize these speculations for what they are. They are not physics but, in the strictest sense, metaphysics. There is no purely scientific reason to believe in an ensemble of universes. By construction these other worlds are unknowable by us. A possible explanation of equal intellectual respectability—and to my mind greater economy and elegance—would be that this is the way it is because it is the creation of the will of a Creator who purposes that it should be so.

The world that science describes seems to me, with its order, intelligibility, potentiality, and tightly knit character, to be one that is consonant with the idea that it is the expression of the will of a Creator, subtle, patient, and content to achieve his purposes by the slow unfolding of process inherent in those laws of nature which, in their regularity, are but the pale reflections of his abiding faithfulness. Yet the eventual futility of the physical universe shows also that the ultimate fulfillment of the Creator's eternal purposes will have to take place beyond this present world—which is what I take to be the meaning of the Christian doctrines of the resurrection of the body and the life of the world to come.

We have been thinking of natural theology very much from the point of view of its relation to the scientific picture of the universe. Yet when most people scan the world for signs of God, it is not to its scientific orderliness that they look. Rather, it is such matters as the incidence of debilitating and destructive disease that concern them. The randomly imposed burdens of unmerited suffering seem to many to call in question assertions that the world is in the care of a loving

God. It is far beyond the scope of this book to explore with any adequacy that agonizing question. At the deepest level I believe that the only possible answer is to be found in the darkness and dereliction of the cross, where Christianity asserts that in that lonely figure hanging there we see God himself opening his arms to embrace the bitterness of the strange world he has made.[20] Science has, however, one small insight to offer. It suggests that the suffering of the world may not be wholly gratuitous. Most of us feel that had we been in charge of creation, we would have managed it better. We would have retained the good and eliminated the bad. The balance and relatedness that the anthropic principle discerns in the structure of a world capable of bringing life into existence shows that it is not as easy as we might have supposed to make such changes, if the universe is to remain subjected to underlying lawful regularity. Of course, it would be possible to conceive of a world in which God intervened on every occasion cancer cells formed in a body so as to eliminate the disease by direct action. Yet it may be that the rationality and faithfulness of God mean that it is only according to his nature to create a world whose lawful regularity reflects these aspects of his character. In that case, it may well be that any such world must contain within it possibilities for what we call good and what we call bad, especially if respect for its given independence means that its evolution must take place through the interplay of chance and necessity. I cannot quite bring myself to echo Leibniz's blithe belief that this is the best of all possible worlds, but the idea is not as absurd as one might have thought.

There is a further difficulty for the Christian who has set before him the hope of an eventual destiny in which God "will wipe away every tear from their eyes, and death shall be no more, neither shall there be mourning nor crying nor pain any more, for the former things have passed away."[21] If such things can ever be, why should they not be now? An answer might lie in the patience of a God content to achieve his purposes through the unfolding of process. It is possible that Love can work only in such a way, out of respect for the beloved.

Natural theology, as an insightful discipline, seems to me to be not only possible but necessary. It is a rational response to the strange and beautiful world that science discloses, with its feel of "more than meets the eye." It is also a healthy corrective to distorting tendencies in other branches of theological thought. While no theological statement would be satisfactory which did not take full account of the human condition, it is also necessary to recognize that there is more to God than his dealings with men. After Job's anguished suffering, the answer that he received from God was to be pointed to the grandeur of creation: "Where were you when I laid the foundation of the earth? Tell me, if you have understanding . . . Have you entered into the springs of the sea, or walked in the recesses of the deep?"[22]

Natural theology counters the view of Feuerbach and his successors that religion is just anthropological projection and the view of Durkheim and his successors that it is a socially induced phenomenon. There are reasons for belief in God lying wholly outside the human world. Although religion has an inescapable interior element, it has an exterior element also. Natural theology protects us from undue emphasis on interiorization. The existential tradition in theology, stemming from Kierkegaard, is not enough.

Finally, we need to take account of the fact that there is a popular alternative natural theology to the one I have been propounding. It is the attempt to assimilate scientific thinking to the patterns of Eastern philosophy rather than Christian thought. Great stress is laid on the elusive dissolving character of quantum physics. A unity realized in ever-changing dynamic process is held to be suggested by the quantum refusal sharply to separate observer and observed, and by fundamental physics' picture of quantum fields in mutual interaction. Such a cosmic web of insubstantiality is congenial to the Eastern mind. Fritjof Capra's book *The Tao of Physics* is a fluent presentation of the thesis that occidental science and oriental thought are basically at one. He writes about his book that it "aims at improving the image of science by showing that there is an essential harmony between the spirit of Eastern wisdom and Western science."[23] I do not find this approach convincing. It seems to me to be lopsided, both in its science and its theology. Theologically, the stress is on mystical experience which though present, as Capra acknowledges,[24] in all religious traditions, is particularly emphasized in the religions of the East. Mystics attain unity with the ground of all being and thus experience God in his immanence. The religions of the West, to a greater extent than those of the East, seek to hold this in tension with experience of God in his transcendence, his kingly rule of the world. The reflection of that in natural theology I believe to be found in aspects of the scientific view to which the orientally inclined seem not to give due attention. One mark of divine transcendence visible in the scientific realm is the lawful intelligibility of the world. I have argued that this is the guarantee of the world's reality over against us (p. 56). All does not dissolve, and the clarity of mathematical form lies at the heart of fundamental physics. I believe that the rationality of the world is more justly recognized in a natural theology of the Christian type than in that of an Eastern type. That is also true of those insights of potentiality and tight structure which speak to the Christian of God's creative purpose in the world. They seem to me to get scant attention from the protagonists of Eastern thought. (However, I must acknowledge that my principal reasons for being a Christian rather than, say, a Buddhist, lie elsewhere than in natural theology.[25])

Modes of Thought

Because science and theology are both seeking understanding of the one reality of the world, they are capable of influencing each other by analogies of thought and practice. The scientist learns certain lessons which might also be relevant to the theologian. One is to be open to the unexpected. We have already noted that the scientific view of the world is full of surprises. If we cannot foretell what we shall find when we enter a new physical regime, then it is to be anticipated that our encounter with God will not always accord with prior expectation. There is a school of thought in modern theology which seeks to contain its understanding within the mental straitjacket of what is considered conceivable "in a scientific age." Such an attitude strikes me as profoundly unscientific. A scientific approach would always seek first to accept and evaluate the phenomena, whatever they might be, and allow experience rather than so-called "reason" (which here, in fact, is often a euphemism for paucity of imagination) to set the agenda.

Following such a procedure can sometimes lead science into situations which seem paradoxical. At the beginning of the nineteenth century, Thomas Young gave convincing demonstration of the wave-like character of light. Eventually, Clerk Maxwell was able to identify the waves as being of electromagnetic origin. It was a splendid achievement, perhaps the greatest of nineteenth-century physics. Yet, as the twentieth century began, the work of Planck and Einstein showed equally indubitably that light behaved as if made up of tiny particles.[26] This clash of wave/particle natures seemed absolutely irreconcilable. No progress would have been made by denying half the evidence. Young and Maxwell, Planck and Einstein, all deserved their due. In the end, the matter was happily resolved by the work of a young Cambridge theoretical physicist, Paul Dirac, whose invention of quantum field theory achieved the counterintuitive feat of combining wave and particle without a taint of paradox.

It is the sign of a mature subject to be able to be true to experience, however hard that experience may be to understand. Better a confused state of loyalty to the facts than a tidy theory obtained by Procrustean oversimplification. One cannot tell the wave-particle story of quantum physics without thinking of the God-man duality of Christ. If Christian experience finds in Jesus elements both human and divine, as I believe it does, then it must hold fast to that experience whatever the intellectual problems involved. We live in a subtle world, and both science and theology need to be subtle in their accounts of it.

One might see Dirac's quantum field theory as a scientific gift to the imagination, a powerful example of the reconciliation of opposites, an occidental realization of the oriental synthesis of yin and yang. There are a number of other scientific notions which seem to me to be potentially fruitful in this parabolic role. One is provided by the hologram. Most of us are familiar with these little plates created by the interference of a beam of laser light reflected from an object with another coherent beam of the same light. They are capable, on reillumination, of recreating a three-dimensional image of the original object. The information about that object is encoded in the hologram in a subtle way. It is totally unlike a photographic negative, where each spot of the object corresponds to a spot on the film. In a hologram, information about the whole object is stored in every spot of the plate, so that if the hologram is cut in two and then reilluminated we obtain, not an image of half the object but a somewhat coarser image of the whole. David Bohm has written of this enfolding of the whole into every part as the presence of an "implicate order."[27] He sees it as conveying the potentiality for a holistic account of the world that conventional science fragments. One thinks of Blake's seeing "a World in a Grain of Sand" or Julian of Norwich being shown "a little thing the size of a hazelnut" and being told that it is "all that is made." Theologians might well feel that holograms provide a useful metaphor for coinherence (the mutuality of humankind) or sacramental presence.

A basic mathematical notion is that of a one-one mapping from one space to another, a correspondence which faithfully reproduces structure. In this way one can establish isomorphism, the essential mathematical equivalence of apparently different realizations of a common structure. Such mappings can change detail whilst preserving form, so that identity is conserved through the unchanging character of what mathematicians call invariants—essential continuity in the presence of inessential change. An example is provided by the topological transformations of curves. Consider a curve drawn on a sheet of rubber. As the rubber is distorted, the curve changes shape, but certain characteristic properties are retained. For example, if the original curve did not cross itself (it is like an oval rather than a figure eight), then this will be true of all the subsequent curves into which it is transformed by the distortions. Here we can see a possible faint metaphor for the preservation of personality in the resurrection transformation from this world to the world to come. Moreover, our topological transformations can be used to turn an "imperfect" oval into a "perfect" (highly symmetric) circle, a metaphor with redemptive overtones.

These ideas are but the toys of thought. Yet story and parable have an important role in theological activity, and the tales of science are well worth listening to for their possible resonance with deeper matters.

6

Levels of Description

The knowledge we obtain from our exploration of the world can be organized into a hierarchy, corresponding to the complexity of the systems treated as basic: physics, chemistry, biochemistry, biology, psychology, sociology, theology. In this chapter we must address ourselves to the question of how these different levels of description relate to each other. The thorough-going reductionist offers us an answer: ultimately all is physics.[1] Everything else is nothing but an epiphenomenal ripple on the surface of a physical substrate, just as waves generated by the wind in a field of corn are nothing but the motions of many ears of wheat. It is an answer of great simplicity and implausibility. Is a Rembrandt self-portrait nothing but a collection of specks of paint? A Shakespearean sonnet nothing but a pattern of ink marks on a sheet of paper? If we take the picture or the printed poem apart that is all we shall find. There will not be an extra ingredient, the spirit of art or poetry, to be found in the residue. It seems to me that we can accept a structural reductionism, that the units out of which all the entities of the physical world are constructed are just the elementary particles studied by fundamental physics. In biology vitalism seems justly dead. The successes of molecular biology do not encourage us to believe that a mysterious entelechy or élan vital is an additive necessary to turn inanimate matter into living being. It is an entirely different proposition to add to this, as our thorough-going reductionist does, a conceptual reductionism which denies the emergence, with increasing complexity of organization, of totally new levels of meaning and possibility which are not in principle reducible to those which lie below them. If structural reductionism is not to lead to conceptual reductionism, a careful investigation is necessary into the extent to which genuine novelty can come into being with these increasing degrees of organization. To what extent is the whole more than the sum of its parts? It must surely be through the enhancing effect of mutual cooperative interaction, made possible by integrating components into a larger unity.

I believe that those who defend the autonomy of their subject against the imperialist claims of physics are right to do so. Animals are made of atoms, but that does not imply that biology is just a complicated corollary to atomic physics. Characteristically, biological concepts need for their understanding the total living setting in which they find their expression. Consider a sentence like: molecular biology has given us considerable biochemical insight into how the genetic blueprint is encoded onto DNA and how messenger RNA transfers appropriate parts of that plan to control the production of proteins. It speaks a hopelessly mixed language. Biochemistry can talk about the molecular dynamics of aggregating amino acids to form proteins. The information-carrying language of "blueprint" and "plan" refers to a different and disjoint type of discourse which only begins to make sense in a cellular context. Arthur Peacocke has written:

In no way can the concept of "information," the *concept* of conveying a message, be articulated in terms of the *concepts* of physics and chemistry, even though the latter can be shown to explain how the molecular machinery (DNA, RNA and protein) operates to carry information. For the concept of information is meaningless except with reference to the functioning of the whole cell, itself conceived in relation to genetics and evolutionary history . . . [Information] is a concept applicable to the molecular system of

DNA, RNA and protein only when these complex structures are linked in networks of interrelations constituting the whole cell.[2]

In other words, it is the wider context that confers the higher meaning. Writing about machines, the psychologist Richard Gregory wrote:

It is silly to expect part-whole descriptions to work when the concepts for describing the whole system are not expressed when the parts are described . . . It is ridiculous to describe a rod of Invar steel lying in a box as a pendulum (it could be used for all manner of things), though it is a pendulum when freely swinging . . . So describing how clocks are *made* is not a good way of describing how they *work*.[3]

For the steel rod, or for the DNA, it is the context which alone furnishes the full meaning.

If reductionism is not the whole story for engineering, it will surely not be the whole story for biology. Yet it is to physics that we shall turn for our exploration of emerging conceptual novelty. Even that most basic of subjects is not monolithic in its structure. We can discern within it a multileveled reality.

It is a statement too trivial to be interesting to say that with a system of only one atom, we should not have the concept of interatomic forces. It is scarcely surprising that when we have several atoms, there are forces between them, even if the details of molecular bonding make theoretical chemistry a more subtle subject than one might have imagined. However, when many atoms are put together to form the regular array of a crystalline solid, then properties emerge which have the feel of genuine novelty. Electrons moving in such an environment are only capable of states of motion with energies lying in certain ranges. The other ranges of energy are inaccessible to them. This band structure, as it is called, is contrary to the way electrons behave in simpler settings. It is also the source of the familiar electrical properties of conductors and insulators. Band structure is an example of one of the many properties of matter which only acquire a meaning when a certain level of organization is reached. The wetness of water is another example. No doubt it is a property of large aggregates of H_2O molecules in appropriate circumstances, and it is certainly not obtained by adding a mysterious wetness ingredient to them. Yet it is conceptually independent of anything one could say about one or a few individual specimens of H_2O.

The most striking level shift in physics is concerned with the interpretation of quantum mechanics. We have discussed the problems involved in trying to understand the quantum mechanical act of measurement (pp. 57–60). How does the reliable everyday world of the laboratory manage to yield a definite result each time it investigates the fitful indeterminate world of quantum particles? We inclined to the view that it is the "large" systems of the measuring apparatus which play this objectifying role (which on a given occasion nail the electron down to being "here" rather than "there" when we address to it the experimental question, "Where are you?"). But these instruments are themselves composed of quantum mechanical components. They are made of elementary particles like everything else. Out of the indeterminate comes the determinator—as conceptually irreducible an emergence, it seems to me, as that of life from inanimate matter, or of freely choosing and responsible persons from their animal ancestors. We do not understand how this level shift takes place in quantum physics; and so, without in any way granting a license for intellectual laziness, it is scarcely surprising that those other level shifts are also beyond our present comprehension. That is, of course, no reason either for denying their existence or for not continuing to seek to understand them.

Niels Bohr suggested, in fact, that there is an idea from quantum theory which might be useful in thinking about such matters as the relation of living beings to their inanimate components.[4] He had earlier introduced the notion of complementarity into atomic physics.

It arises from the partial knowledge which is all that Heisenberg's uncertainty principle permits us to obtain about a physical system. We can know either where the electron is or what it is doing but not both. As a consequence there is one way of describing its quantum mechanical behavior which is phrased wholly in terms of position and another description phrased wholly in terms of momentum. Each of these descriptions is complete in itself. It says all that can be said. They exclude each other, in the sense that if we adopt one, the other is ruled out of court. This corresponds to the experimental fact that the use of apparatus to determine position (where it is) is incompatible with the use of apparatus to determine momentum (what it is doing). In a word, they are complementary ways of looking at the same physical reality.

Bohr suggested that the idea might prove of wider applicability. In particular, he proposed that there might be a complementarity between life and atomic physics. Certainly if I break up a living creature into its constituent parts I kill it. The deficiency in Bohr's suggestion is that it is really little more than a rephrasing of that last sentence. As such, it does not convey new understanding. In quantum theory the relation between the two complementary descriptions of position and momentum is well understood, and we know how to transform from one picture into the other. That is not the case with life and atomic physics. Yet Bohr's remark is valuable insofar as it lays equal emphasis on the whole (life) as well as the parts (atoms). Holism is as significant as atomism.

The motto of science has been "Divide and rule"; its methodology principally reductionist in technique. It is important not to let the dazzling successes of that approach blind us to the need to give proper attention to holistic ideas. It is interesting that on both the frontiers of physics, the very small and the very large, we are reminded of this need.

Einstein and his collaborators Podolsky and Rosen drew attention in 1935 to a remarkable property of quantum theory. It is sometimes called the EPR experiment. It states that once two quantum particles have interacted with each other, they retain the power of mutual influence however widely they may subsequently separate.[5] For example, if two electrons have been in interaction and I then investigate one of them here in the laboratory, that investigation will have an instantaneous effect upon the other, even if it is "beyond the moon" (as we conventionally say to signify great distance). Einstein thought this so counterintuitive that he believed it indicated an incompleteness in the conventional quantum theory that he so greatly distrusted. However, most of us do not see it that way. Instead, for us the EPR experiment indicates an astonishing togetherness in separation, even for elementary particles. This view has received excellent empirical confirmation. At the very root of reductionist physics, we find holism reasserting itself. Even electrons are not self-contained solitaries. The thought is reinforced by the further consideration that they are all excitations in a single quantum field and that the Pauli exclusion principle (which states that the presence of an electron in a given state of motion prevents any other electron entering that state of motion) also links the behavior of one electron with that of all others. Margenau went so far as to refer to the latter phenomenon as electrons being subject to a law "which regulated their social behaviour."[6]

Most of us have seen, in a science museum or elsewhere, a Foucault pendulum. As it swings for many hours, its plane of oscillation slowly rotates. The effect is due to the rotation of the Earth. A calculation based on Newtonian mechanics enables us to disentangle from the behavior of the pendulum the period of the Earth's rotation which is involved. The answer comes out at about 23 hours, 56 minutes. This is also the length of the sidereal day, the time it takes for the Earth to rotate on its axis with respect to the fixed stars. (It is a little bit shorter than the familiar 24 hours between noon and noon because the addition of the Earth's motion in its orbit round the sun means that it takes that extra four minutes to get the sun back into position relative to the Earth. It has got a little left behind, so to speak. The fixed stars are too far away to be affected by the orbital motion.) A very remarkable coincidence is involved in this result. Foucault's pendulum is measuring the period of the Earth's rotation with respect to what physicists call an inertial frame of reference. This is a standard situation in which the laws of dynamics take their simplest form. The equivalence of the Foucault period with the sidereal day means that this standard reference frame is at rest with respect to the fixed stars, that is the surrounding matter of our galaxy. Never mind if you have found that a little difficult to follow. The essential point is this: that the local dynamical properties here on Earth appear to be determined in some of their details by the overall distribution of stars in the galaxy. The significance of this was first pointed out by Ernst Mach, and the idea that local dynamics (physics here) is influenced by global properties of the universe (matter everywhere) is called Mach's principle. You could call it scientific astrology. It is a remarkable cosmological example of a holistic effect; the behavior of a part is determined by the whole.

Even our experience of time may be in some sense a holistic phenomenon. One of its most striking features is its asymmetry; we can remember the past, but we cannot foretell the future. What makes this puzzling is that the basic laws of physics, the reductionist equations, seem almost perfectly symmetrical between past and future. (There is a small degree to which they are not, involving the weak nuclear forces, but this is not relevant to our present discussion.[7]) A film of two elementary particles interacting (were it possible to make one) would have no intrinsic time direction; it would make sense run either forwards or backwards. That is not so for us. The film in which we grow younger is clearly reversed. The origin of time's direction is not well understood or agreed among physicists, but it seems to be connected with the operation of the second law of thermodynamics. This associates the direction of time with the direction of increasing entropy or state of disorder. The television advertisement in which the higgledy-piggledy pile of sweets jumps up and rearranges itself tidily in the box is certainly being run backwards, because it opposes this trend from order to disorder. Thus understood, the direction of time's arrow is a property of the macroscopic whole not found in the microscopic parts. Its irreversibility may also provide an important clue to how it comes about that macroscopic objects determine the outcome of microscopic quantum events.

The most important and perplexing problem in this general area of level relationships is the perpetual puzzle of the connection of mind and brain. To the thorough-going reductionist the answer is easy: mind is the epiphenomenon of brain, a mere symptom of its physical activity. This is a highly curious point of view to adopt. The most basic experience we have is mental, that stream of consciousness which constitutes what is sometimes called our I-story. The world which we believe we perceive around us, and which is the subject of what we may call the it-story, is itself a construct of the material furnished by the I-story. Among its objects is our brain with its elaborate electrochemical functioning. You will realize from the robustly realist view I take of science that I am not about to retreat with Bishop Berkeley into the idealist castle. I do not doubt that the it-story is a veridical construction. But to use it to cast doubt on the autonomous validity of the I-story is to saw off the epistemic branch on which we are resting.

Of course, in practice none of us does that. Commenting on nineteenth-century materialism, Mascall dryly observes, "However sure the scientist might be that other people were only elaborate machines, his protocol contained an escape clause for himself."[8] The fallacy of behaviorism, the theory that people are to be considered solely in terms of the it-story of their external reactions, without any empathetic attempt to comprehend the I-story that lies within, is that we cannot treat ourselves that way, as mere responders to the positive and negative feedback of our environment. We *know* that is wrong, as surely as we know anything.

The reductionist program in the end subverts itself. Ultimately, it is suicidal. Not only does it relegate our experiences of beauty, moral obligation, and religious encounter to the epiphenomenal scrapheap. It also destroys rationality. Thought is replaced by electro-chemical neural events. Two such events cannot confront each other in rational discourse. They are neither right nor wrong. They simply happen. If our mental life is nothing but the humming activity of an immensely complexly connected computerlike brain, who is to say whether the program running on the intricate machine is correct or not? Conceivably some of that program is conveyed from generation to generation via encoding in DNA, but that might still be merely the propagation of error. If we are caught in the reductionist trap, we have no means of judging intellectual truth. The very assertions of the reductionist himself are nothing but blips in the neural network of his brain. The world of rational discourse dissolves into the absurd chatter of firing synapses. Quite frankly, that cannot be right and none of us believes it to be so.

The I-story is to be taken seriously, but how it interlocks with the it-story is a mystery even more profound than the problem of how dependable measuring apparatus interlocks with the fitful world of quantum physics. Alan Turing in a celebrated article posed the question, "Can machines think?"[9] He suggested that it should be answered affirmatively if a person, in a conversation conducted via a teleprinter link, could not tell whether his addressee was a machine or another person. The Turing test is framed in the behaviorist spirit of attention to external reaction. It would be truly convincing only if a sustained conversation led to the conviction that an I-story lay behind the responses. No doubt a computer could be programmed to simulate that in a limited exchange. Whether it could do so in an acquaintance-ship lasting many months or years is another question. An essential ingredient in our experience of other self-conscious beings is the sign they give of freely exercised independence, the creatively open nature of their response. No account of the relation of mind and brain, the I- and the it-stories, will be adequate which does not do justice to this impression, which is reinforced by our own internal experience of choice and responsibility. Equally, any such account must do justice to the influence of physical circumstance on mental behavior, which the effects of drugs and brain damage make only too clear. The reconciliation of these two requirements is not easy. Despite the creakiness introduced by quantum theory, the physical system of the brain's neural network looks a pretty determinate structure in its operation. How can this leave room for the experienced freedom of the mind? Donald Mackay has discussed the problem.[10] For argument's sake he is prepared to concede the possibility of a determinate brain in which every mental state has a corresponding physical image. In principle, an external observer could then investigate your present brain state and predict your future thoughts and behavior. If you were offered such a prediction you might falsify it, but that itself does not dispose of the deterministic picture since that act could be interpreted as the reaction of your brain to an external influence not taken into account in making the original calculation, namely the proffered prediction. We may suppose that a super-scientist predictor could allow for such a perturbation and offer you an improved *self-consistent* prediction which took its own influence correctly into account. Would this not show that what you thought was choice was an unfolding of inevitability? Mackay thinks not. The prediction has been made on the basis that you believe it when offered. Speaking in the character of the one subjected to the procedure, Mackay says of the prediction that it has claim to his assent and so is a story he would not be mistaken to believe. Nevertheless,

it would become accurate *if and only if* I believed it. If I did not believe then my brain would not be in the state supposed by Brain Story 2 [the self-consistent prediction]. So his cooked-up story is also a story I would not be mistaken to *disbelieve*.

He goes on to say:

Even on the strongest mechanistic assumptions, there cannot exist a complete specification of the immediate future of my brain with an unconditional claim to my assent—such that I would be correct to believe and mistaken to disbelieve it if only I knew it. In that sense, even the doctrine of physical determinism would not imply my future is inevitable for me.[11]

I do not see that. On the basis stated, the question of whether you believe the prediction when offered has itself a determinate answer, for it involves an action of your brain. This fact should, of course, be part of the self-consistent calculation. If, say, the answer is that you will believe, then the apparently falsifying possibility that you will disbelieve is hypothetical only and incapable of actual realization. If, however, the answer is that you will disbelieve, then that is the appropriate brain state figuring in the calculation, whose consequences will inevitably unroll themselves, your disbelief notwithstanding. In this case it would be your belief which, paradoxically, would upset the subtle calculations, but belief is an option inaccessible to you in the circumstances. In no way does it seem that the inevitability of the future is mitigated, or freedom of the mind preserved.

There are, anyway, a number of dubious points about the whole procedure. It is not clear that an external investigator could obtain the fantastically detailed knowledge of your brain state needed to make the predictions, without grave perturbation, even destruction, of your brain. (This is the Bohr complementarity point.) It is possible also that the self-consistency of the calculations is more difficult to achieve than we have been supposing. Self-referring operations (recursions, as the mathematicians say) are notoriously tricky. We can see that even in logic, with self-referring propositions, such as "This sentence is false." If it is true it is false and if it is false it is true. (Perhaps offering Donald Mackay a prediction of his behavior leads to a similar dilemma?) Paul Davies has given a pretty little physical parable of the surprising effects of self-reference in terms of the Mobius strip.[12] This is a ribbon of paper joined onto itself with one twist. Locally (that is, in any small region) it appears to be two-sided, but actually it is only one-sided, in the sense that you can reach the point apparently on the "other side" by taking a trip once along the whole length of the strip. (Try it, if you don't believe me!) This unexpected global (holistic) property of one-sidedness is due to the recursion of the twisted self-join.

Mackay's discussion is externalized, almost behaviorist in tone; it lays too much stress on the it-story aspect of external prediction. The recursive complexity is greatly increased if we make the necessary transposition into the I-story. Self-consciousness is by definition self-referring. John Locke wrote concerning what it is to be a person, that it is to be

a thinking intelligent being, that has reason and reflection, and can consider itself as itself, the same thinking thing, in different times and places: which it does only by the consciousness which is inseparable from thinking, and it seems to me essential to it; for it is impossible for anyone to perceive without perceiving that he does perceive.[13]

The way that autonomous mind is sustained by the physical operation of the brain must be supposed to involve many levels recursively linked, in which the mental and the physical are subtly interwoven.[14] As a very simple example consider the following suggestion by Don Cupitt.[15] Let us suppose that the physical correlate of seeing is activity in a particular part of the cortex. By trepanning and a system of mirrors and microscopes, one can suppose oneself put in the position of being able to observe that activity. On a reductionist basis the thing observed and the observing of it would be identical, which leads to all sorts of paradox. As Cupitt says, "How can a physical event *be* its own watching of itself occurring?" Freed from the reductionist trap, we can see this as an elementary but striking example of a recursive relationship involving the mental and physical levels of experience.

Of course, I do not pretend that such considerations can, in our present state of knowledge, do more than provide some mildly helpful hints about the relationship of mind and brain. At best they constitute a sort of scientific poetry. (Perhaps I should say verse.) Until we know better how to integrate them, let us at least hold fast to our basic personal experience of choice and responsibility without denying the neurological insight that our mental activity is incarnated in our brains. These are complementary aspects of the whole person, just as wave and particle are complementary aspects of light. And, at the risk of being tedious, let me say once more that the emergence of mind from matter is only a degree more mysterious than the emergence of objectifying measuring instruments from the fitful quantum world.

7

One World

It was said, perhaps unjustly, of the great nineteenth-century experimental physicist Michael Faraday, who was a committed Christian believer, that when he went into his laboratory, he forgot his religion, and when he came out again, he forgot his science. I hope that is not true. We live in one world, and science and theology explore different aspects of it. The burden of our tale has been that, despite the obvious differences of subject matter, the two disciplines have in common the fact that they both involve corrigible attempts to understand experience. They are both concerned with exploring, and submitting to, the way things are. Because of this, they are capable of interacting with each other: theology explaining the source of the rational order and structure which science both assumes and confirms in its investigation of the world; science by its study of creation setting conditions of consonance which must be satisfied by any account of the Creator and his activity. Their relationship is not free from puzzles, but I have sought to show that no act of mental compartmentalism or dishonest adjustment is required of those who take with equal seriousness the stories told by science and by faith.

Reality is a multi-layered unity. I can perceive another person as an aggregation of atoms, an open biochemical system in interaction with the environment, a specimen of *homo sapiens*, an object of beauty, someone whose needs deserve my respect and compassion, a

brother or sister for whom Christ died. All are true and all mysteriously coinhere in that one person. To deny one of these levels is to
diminish both that person and myself, the perceiver; to do less than
justice to the richness of reality. Part of the case for theism is that in
God the Creator, the ground of all that is, these different levels find
their lodging and their guarantee. He is the source of connection, the
one whose creative act holds in one the worldviews of science, aes-
thetics, ethics and religion, as expressions of his reason, joy, will, and
presence. This interlocking character of the world of creation finds its
fullest expression in the concept of sacrament, an outward and visible sign of an inward and spiritual grace, a wonderful fusion of the
concerns of science and theology. Thus in the Eucharist, bread and
wine which, in the words of the liturgy, "earth has given and human
hands have made," become the body and blood of Christ, the source
of spiritual life. The greatest sacrament, compared to which all others
are types and shadows, is the Incarnation in which "the Word became
flesh and dwelt among us, full of grace and truth; we have beheld his
glory, glory as of the only Son from the Father."[1]

The Word, the *Logos,* combines two notions, one Greek, one
Hebrew. For the Greek the *logos* was the rational ordering principle of
the universe. For the Hebrew the word of the Lord was God's activity
in the world. (In Hebrew *ddbar* means both word and deed; Hebrew
is a language based on verbs, on action.) Science discerns a world of
rational order developing through the unfolding of process, a fusion
of Greek and Hebrew insights. Theology declares that world in its
scientific character to be an expression of the Word of God. For "all
things were made through him, and without him was not anything
made that was made."[2]

Notes

Chapter 1. The Post-Enlightenment World

1. For example, M. B. Foster, "The Christian Doctrine of Creation and the Rise of Modern Natural Science" in *Mind* 43 (1934): 446.

2. W. Temple, *Nature, Man and God* (London: Macmillan, 1935), 57.

3. Quoted in J. Wesley's *Works* (Grand Rapids: Zondervan, 1958–59), 13:449.

4. W. Blake, "A Vision of the Last Judgement" in *Descriptive Catalogue* (1810).

Chapter 2. The Nature of Science

1. K. Popper, *Conjectures and Refutations* (London: Routledge and Kegan Paul, 1969), 216.

2. For an account of the development of elementary particle physics, see J. C. Polkinghorne, *The Particle Play* (New York: W. H. Freeman, 1979).

3. N. R. Hanson, *Perception and Discovery* (San Francisco: Freeman Cooper, 1969), ch. 9.

4. For details of these matters, see J. C. Polkinghorne, *The Quantum World* (London: Longman, 1984), ch. 5.

5. M. Polanyi, *Personal Knowledge* (London: Routledge and Kegan Paul, 1958).

6. T. Kuhn, *The Structure of Scientific Revolutions,* 2nd ed. (Chicago: Chicago University Press, 1970).

7. Ibid., 94.

8. P. Feyerabend, *Against Method* (London: Verso, 1975), 23.

9. Ibid., 215.

10. A. Pickering, *Constructing Quarks* (Edinburgh: Edinburgh University Press, 1984), 406.

11. W. H. Newton-Smith, *The Rationality of Science* (London: Routledge and Kegan Paul, 1981), 14.

12. Popper, *Conjectures and Refutations*, 55.

13. K. Popper, *The Logic of Scientific Discovery* (London: Hutchinson, 1968), 111.

118 NOTES

14. Newton-Smith, *Rationality of Science*, 62.
15. See, for example, ibid., ch. 2.
16. B. d'Espagnat, *The Conceptual Foundations of Quantum Mechanics* (Menlo Park, CA: Benjamin, 1971), 474.
17. H. Margenau, *The Nature of Physical Reality* (New York: McGraw-Hill, 1950), 295.
18. A. S. Eddington, *Fundamental Theory* (Cambridge: Cambridge University Press, 1946).
19. Newton-Smith, *Rationality of Science*, 209.
20. J. R. Carnes, *Axiomatics and Dogmatics* (New York: Oxford University Press, 1982), 14.
21. For a brilliantly lively account, see D. Hofstadter, *Gödel, Escher, Bach* (Harlow, UK: Harvester, 1979).
22. Ibid., 19.

Chapter 3. The Nature of Theology

1. P. Davies, *God and the New Physics* (London: Dent, 1983), 6.
2. Ibid., 220.
3. Deut. 6:16.
4. Job 13:15 (KJV).
5. A. N. Whitehead, *Religion in the Making* (New York: MacMillan, 1926), 57.
6. Davies, *God and the New Physics*, viii.
7. W. James, *The Varieties of Religious Experience* (London: Collins, 1977), 404.
8. A. Hardy, *The Spiritual Nature of Man* (New York: Oxford University Press, 1979).
9. Ibid., 1.
10. See J. C. Polkinghorne, *The Way The World Is* (London: Triangle, 1983).
11. I. G. Barbour, *Issues in Science and Religion* (London: SCM, 1966), 226.
12. For details, see J. C. Polkinghorne, *The Quantum World* (London: Longman, 1984), 58–59.
13. 1 Tim. 6:15–16.
14. Ps. 113:7.
15. These ideas are beautifully worked out in W. H. Vanstone, *Love's Endeavour, Love's Expense* (London: Darton, Longman and Todd, 1977).
16. Augustine, *Confessions*, XI.
17. B. Pascal, *Pensées* (New York: Penguin, 1966), 95.
18. Phil. 2:12–13.
19. D. M. Baillie believed this to be the fundamental Christian paradox and upon it based his christology: *God Was In Christ* (London: Faber, 1956).
20. J. R. Carnes, *Axiomatics and Dogmatics* (New York: Oxford University Press, 1982), 68.

21. Ibid., ch. 5.

22. J. C. Polkinghorne, *The Particle Play* (New York: W. H. Freeman, 1979), ch. 5.

23. Quoted in Barbour, *Issues in Science and Religion*, 249.

24. Carnes, *Axiomatics and Dogmatics*, ch. 7.

25. J. Hick, *Faith and Knowledge* (Ithaca, NY: Cornell University Press, 1966), ch. 8.

26. 1 Cor. 13:12.

27. For example, see C. Bryant, *Jung and the Christian Way* (London: Darton, Longman and Todd, 1983).

28. D. Cupitt, *The Sea of Faith* (London: BBC Publications, 1984), 270.

29. R. B. Braithwaite, *An Empiricist's View of the Nature of Religious Belief* (Cambridge: Cambridge University Press, 1955).

30. For example, see A. E. Harvey, ed., *God Incarnate—Story and Belief* (London: SPCK, 1981).

Chapter 4. The Nature of the Physical World

1. J. Monod, *Chance and Necessity* (London: E. T. Collins, 1972).

2. T. de Chardin, *The Phenomenon of Man* (London: E. T. Collins, 1959).

3. See J. C. Polkinghorne, *The Quantum World* (London: Longman, 1984), especially chs. 4–6.

4. Quoted in M. Jammer, *The Philosophy of Quantum Mechanics* (New York: Wiley, 1974), 204.

5. See J. C. Polkinghorne, *The Particle Play* (New York: W. H. Freeman, 1979), chs. 4 and 7.

6. See Polkinghorne, *Quantum World*, ch. 8.

7. W. Heisenberg, *Physics and Philosophy* (London: Allen and Unwin, 1955), 160.

8. Quoted in P. Davies, *God and the New Physics* (London: Dent, 1983), 221.

9. A. Pickering, *Constructing Quarks* (Edinburgh: Edinburgh University Press, 1984), 413.

10. E. L. Mascall, *Christian Theology and Natural Science* (New York: Longman, 1956), 175.

11. See Polkinghorne, *Quantum World*, ch. 6.

12. Ibid., 68.

13. Monod, *Chance and Necessity.*

14. F. Hoyle, *The Intelligent Universe* (London: Michael Joseph, 1983), 12.

15. F. H. Crick, *Life Itself* (London: Futura, 1982).

16. Quoted in A. R. Peacocke, *Creation and the World of Science* (Oxford: Oxford University Press, 1979), 165.

17. Quoted in I. G. Barbour, *Issues in Science and Religion* (London: SCM, 1966), 92.

18. Monod, *Chance and Necessity*, 10.

19. Ibid., 167.

20. Peacocke, *Creation and the World of Science*, 94.

21. Ibid., 106.

22. Quoted by D. Cupitt, *The Sea of Faith* (London: BBC Publications, 1984), 203.

23. Julian of Norwich, *Revelations of Divine Love* (New York: Penguin, 1966), 80.

24. Ps. 8:3–4.

25. J. A. Baker, *The Foolishness of God* (London: Darton, Longman and Todd, 1978), 53.

26. For example, see S. Weinberg, *The First Three Minutes* (London: Andre Deutsch, 1977).

27. Davies, *God and the New Physics*, 179.

28. For a popular account, see A. H. Guth and P. J. Steinhardt, "The Inflationary Universe" in *Scientific American* (May 1984): 90.

29. J. Macquarrie, *Principles of Christian Theology*, 2nd ed. (London: SCM, 1977), 256.

30. C. A. Coulson, *Science and Religion: A Changing Relationship* (Cambridge: Cambridge University Press, 1955), 2.

31. P. B. Medawar, *Advice to a Young Scientist* (New York: Harper and Row, 1979), 31.

Chapter 5. Points of Interaction

1. Quoted in A. Moszokowski, *Conversations with Einstein* (New York: Horizon, 1970).

2. Quoted in I. G. Barbour, *Issues in Science and Religion* (London: SCM, 1966) , 30.

3. E. Schillebeeckx, *Jesus* (London: E. T. Collins, 1979), 579.

4. P. Davies, *God and the New Physics* (London: Dent, 1983), ix.

5. S. Weinberg, *The First Three Minutes* (London: Andre Deutsch, 1977), 149.

6. Augustine, *Confessions*, V, 3.

7. For a useful account of all issues involving chance, see D. J. Bartholomew, *God of Chance* (London: SCM, 1984).

8. D. M. Mackay, *The Clockwork Image* (London: Inter-Varsity Press, 1974).

9. A. R. Peacocke, *Creation and the World of Science* (Oxford: Oxford University Press, 1979), especially chs. 3 and 5.

10. Bartholomew, *God of Chance*, 81.

11. B. Pascal, *Pensées* (New York: Penguin, 1966), 125.

12. W. G. Pollard, *Chance and Providence* (London: Faber, 1958).

13. Mark 4:37–41.

14. Josh. 10:6–15.

15. Matt. 17:24–27.

16. Quoted in E. L. Mascall, *Christian Theology and Natural Science* (New York: Longman, 1956), 180.

17. For a detailed account of my views on the resurrection, see J. C. Polkinghorne, *The Way The World Is* (London: Triangle, 1983), ch. 8.

18. 1 Cor. 6:14.

19. Davies, *God and the New Physics*, 98.

20. Polkinghorne, *Way The World Is*, ch. 7.

21. Rev. 21:4.

22. Job 38:4, 16.

23. F. Capra, *The Tao of Physics* (Aldershot, UK: Wildwood House, 1975), 25. Capra appealed to bootstrap ideas that have been abandoned by particle physicists (see J. C. Polkinghorne, *The Particle Play* (New York: W. H. Freeman, 1979), ch. 6).

24. Ibid., 17.

25. See Polkinghorne, *Way The World Is*, ch. 10.

26. J. C. Polkinghorne, *The Quantum World* (London: Longman, 1984), ch. 2.

27. D. Bohm, *Wholeness and the Implicate Order* (London: Routledge and Kegan Paul, 1980).

Chapter 6. Levels of Description

1. An elegant account of extreme reductionism is given in P. W. Atkins, *The Creation* (San Francisco: W. H. Freeman, 1981).

2. F. K. Hare, ed., *The Experiment of Life* (Toronto: University of Toronto Press, 1983), 54.

3. R. L. Gregory, *Mind in Science* (London: Weidenfeld and Nicholson, 1981), 93.

4. N. Bohr, *Atomic Physics and Human Knowledge* (New York: Wiley, 1958).

5. J. C. Polkinghorne, *The Quantum World* (London: Longman, 1984), ch. 7.

6. Quoted in I. G. Barbour, *Issues in Science and Religion* (London: SCM, 1966), 296.

7. See J. C. Polkinghorne, *The Particle Play* (New York: W. H. Freeman, 1979), 45f.

8. E. L. Mascall, *Christian Theology and Natural Science* (New York: Longman, 1956), 9.

9. A. M. Turing, "Computing Machines and Intelligence" in *Mind* 59 (1950): 433.

10. D. M. Mackay, *Brains, Machines and Persons* (London: Collins, 1980), ch. 5.

11. Ibid., 95.

12. P. Davies, *God and the New Physics* (London: Dent, 1983), 95.

13. J. Locke, *Essay Concerning Human Understanding*, XXVII, 9.

14. There may be connections here with Gödel's theorem (p. 31); see D. Hofstadter, *Gödel, Escher, Bach* (Harlow, UK: Harvester Press, 1979); D. R. Hofstadter and D. C. Dennett, eds., *The Mind's I*. (Harlow, UK: Harvester, 1981).

15. D. Cupitt, *The Worlds of Science and Religion* (London: Sheldon, 1976), 93–94.

Chapter 7. One World

1. John 1:14.

2. John 1:3.

Glossary

algorithm A specifiable calculational procedure.

alpha particles Helium nuclei.

amino acids Organic molecules which are the building blocks out of which proteins are constructed.

anthropic principle The collection of scientific insights which indicates that a universe capable of evolving systems as complicated as men must have a delicate balance in the structure of its fundamental forces and (perhaps) special initial circumstances.

apophatic theology That theological tradition which acknowledges the otherness and mystery of God, placing him beyond the access of intellectual enquiry.

archetype Powerful symbols whose activity Jung discerned in the unconscious minds of men and women and whose universal character convinced him of the existence of a collective unconscious shared by all.

behaviorism That account of human psychology which speaks only of external behavior interpreted as response to environmental stimuli.

big bang The earliest event in the universe's history accessible to science; that singular moment in which cosmic matter appears to have exploded from a point of infinite compression.

black hole A highly condensed state of matter brought about by collapse under gravitational forces. The resulting gravitational field is so strong that even light cannot escape from it. However, Stephen Hawking has shown that quantum effects (tunneling) permit "black" holes to shine to some extent.

bubble chamber A device for detecting elementary particles which contains a superheated liquid about to boil. The passage of a charged particle leaves a trail of bubbles, thereby manifesting the particle's track.

Cartesianism That philosophical system, originated by René Descartes, which assigns a dual structure to the world in terms of mind and matter.

complementarity The quantum mechanical insight which recognizes that there are alternative, mutually exclusive, modes of description, each complete in itself.

cosmology The study of the structure and history of the universe.

deism That theological viewpoint which sees God as the originator of the world who then leaves it to evolve according to its own laws, taking no further interest in it.

DNA (deoxyribonucleic acid) The highly complex organic macromolecule which is the carrier of the genetic code.

electronic counters Devices for detecting and registering the passage of charged particles.

Enlightenment The eighteenth-century intellectual movement which emphasized a scientific style of rationality and a critical attitude to human knowledge.

EPR experiment The surprising result that quantum particles which have once interacted with each other retain the power of mutual influence however widely they subsequently separate from each other.

existentialism That philosophical attitude which places primary emphasis on being, experienced in authentic human personhood.

gauge theories Highly symmetric and tightly knit specimens of quantum field theories. They play a fundamental role in current thinking about theoretical physics.

Gödel's theorem The mathematical result that systems of sufficient complexity to include the whole numbers always contain propositions which are stateable but not decidable within that system.

graviton The (hypothetical) particle which is thought to be the carrier of the gravitational force in quantum field theory.

hard scattering Collisions at high energy between elementary particles in which large amounts of momentum are transferred between them, thereby producing substantial changes in their states of motion.

hidden variables An interpretation of quantum theory which attributes its statistical character to ignorance about unobservable (hidden) features whose precise character in fact determines completely all that happens.

holism That attitude which places emphasis on wholes rather than the parts of which they are composed.

hologram An interference pattern which on re-illumination yields a three-dimensional image.

idealism That philosophical attitude which attributes reality only to mental phenomena.

induction The philosophical problem of the validity of general conclusions based on the examination of particular instances.

inertia The mechanical property which measures a body's resistance to a change in its state of motion.

inflation A hypothetical stage in the universe's early history in which a very rapid expansion took place (the "boiling" of space-time).

instrumentalism That philosophical attitude which regards the purpose of science to be solely the production of the power to manipulate circumstances to specified ends.

isomorphism A mathematical relationship exhibiting identity of structure between two apparently different systems.

mechanics, classical The study of the motion of bodies according to Newtonian laws.

mechanics, quantum That modification of classical mechanics which is necessary to give an accurate account of the behavior of systems of atomic size or less.

mysticism The unitive experience of God as the ground of all being.

natural theology The attempt to find knowledge of God through the use of reason and the inspection of the world.

neo-Darwinism The account of the evolution of life which results from combining the idea of natural selection through the survival of the fittest with the insights of modern genetics.

neuron(e) A nerve cell in the brain. The average human being has about a hundred thousand million of them.

neutral current A term relating to particle interactions induced by an elusive neutral particle called the neutrino. In *charged current* interactions the neutrino is changed into a readily detectable charged particle, such as an electron. In neutral current interactions, however, the neutrino preserves its identity. This makes the observation of such interactions difficult.

neutron An electrically neutral particle which, with its positively charged partner the proton, forms the constituents of a nucleus.

paradigm An overall scientific worldview.

positivism That philosophical attitude which asserts that science is solely concerned with directly measurable quantities and that its aim is simply

positivism (cont.)
the harmonious reconciliation of observations without attempting to speak of an underlying physical reality.

process theology That theological attitude, based on the thought of A. N. Whitehead, which assigns a role for temporal process in God by which he realizes his "consequent nature" through the evolution of the world.

protein Highly complex molecules, composed of long chains of amino acids, which are essential to all forms of life.

quantum field theory A formalism resulting from applying quantum theory to a field, that is to an entity, like the electromagnetic field, which is extended to all points of space and time. Fields, by their nature, have wavelike properties. The application of quantum mechanics introduces a degree of countability into their structure which leads to a particle interpretation. Thus quantum field theory provides the basis for an understanding of the duality of wave and particle which characterizes the physics of the very small.

quantum logic That modification of the laws of classical logic which is necessary to accommodate the peculiarities of quantum mechanics. It arises because of a characteristic quantum possibility (called "super-position") which allows, for example, a particle to be in a state where it is neither definitely "here" nor definitely "there" but has certain prob-abilities of being found either "here" or "there." This is a middle term of a type undreamed of by Aristotle.

quarks and gluons Particles such as protons and neutrons are now under-stood to be composed of yet more elementary constituents called quarks. The particles which are carriers of the force needed to make the quarks stick together are, I regret to say, called gluons.

realism That philosophical attitude which asserts the reality of a world external to the observer whose detailed structure is open to the investi-gation of science.

reductionism That attitude which speaks of wholes solely in terms of the properties of their parts.

relativity, general Einstein's theory of gravitation which describes effects in terms of the curvature of space-time.

relativity, special That version of mechanics necessary to describe accurately the motion of particles whose velocities are appreciable fractions of the velocity of light.

revelation God's specific acts of self-disclosure.

space-time The four-dimensional mathematical space formed by adding a temporal dimension to the three spatial dimensions. It affords the setting for an integrated account of physical process, and it is a natural construct for expressing special relativity, which closely associates space and time.

superconductivity The phenomenon that certain metals at low temperatures lose all their electrical resistance.

supergravity A very highly symmetrical gauge theory of particles and their interaction with gravity.

synchronicity The influence, posited by Jung, whereby it is suggested that significant coincidences come about.

uncertainty principle Heisenberg's statement that more exact information about where a particle is (position) implies greater ignorance about what it is doing (momentum), and vice versa.

W and Z particles Heavy particles which are carriers of the weak nuclear forces responsible for such processes as certain decays of particles. The Ws are the intermediaries of charged current interactions, and the Zs are intermediaries of the neutral current interactions.

Index